T0335916

Consumer Adoption and Usage of Broadband

Yogesh Kumar Dwivedi
Swansea University, UK

IRM Press
Publisher of innovative scholarly and professional
information technology titles in the cyberage

Hershey • New York

Acquisition Editor:	Kristin Klinger
Senior Managing Editor:	Jennifer Neidig
Managing Editor:	Sara Reed
Development Editor:	Kristin Roth
Copy Editor:	Katie Smalley
Typesetter:	Jeff Ash
Cover Design:	Lisa Tosheff
Printed at:	Yurchak Printing Inc.

Published in the United States of America by
     IRM Press (an imprint of IGI Global)
     701 E. Chocolate Avenue, Suite 200
     Hershey PA 17033-1240
     Tel: 717-533-8845
     Fax: 717-533-8661
     E-mail: cust@igi-pub.com
     Web site: http://www.irm-press.com

and in the United Kingdom by
     IRM Press (an imprint of IGI Global)
     3 Henrietta Street
     Covent Garden
     London WC2E 8LU
     Tel: 44 20 7240 0856
     Fax: 44 20 7379 0609
     Web site: http://www.eurospanonline.com

Library of Congress Cataloging-in-Publication Data

Consumer adoption and usage of broadband / Yogesh K. Dwivedi ... [et al.].
    p. cm.
  Summary: "Even with the pervasiveness of broadband and high-speed communication technologies, many
consumers are not adopting the latest technologies. This book develops a conceptual model for examining
consumer adoption, usage, and impact of broadband utilizing various methodologies, providing a clear win-
dow into the rational decisions of potential broadband consumers"--Provided by publisher.
  Includes bibliographical references and index.
  ISBN 978-1-59904-783-6 (hardcover) -- ISBN 978-1-59904-785-0 (ebook)
  1. Broadband communication systems. 2. Consumer behavior. 3. Consumers' preferences. 4. Telecommunication systems--Government policy. 5. Motivation research (Marketing)--Case studies. I. Dwivedi,
Yogesh K.
  TK5103.4.C66 2008
  384--dc22
               2007007290

British Cataloguing in Publication Data
A Cataloguing in Publication record for this book is available from the British Library.

All work contributed to this book is new, previously-unpublished material. The views expressed in this book
are those of the authors, but not necessarily of the publisher.

Dedicated with love and gratitude to my sister Sudha.

Comsumer Adoption and Usage of Broadband

Table of Contents

Foreword..ix

Preface..xi

Division 1: The United Kingdom Case Study

Section 1.1: The Theoretical Underpinning

Chapter 1
Introduction to Broadband Adoption and Usage Research..........................1
Defining the Research Problem 2
The State of Broadband Adoption, Usage, and Impact Research 6
Research Aims and Objectives..8
Research Approach 10
Research Contributions...11
Summary 12
References 12

Chapter 2
Conceptual Model for Examining Consumer Broadband Adoption,
Usage, and Impact **16**
Technology Diffusion and Adoption Theories................................. 17
Foundations of the Proposed Conceptual Model............................ 23
Description of the Proposed Conceptual Model 25
Attitudinal Constructs.. 29
Normative Constructs 32
Control Constructs 34
Demographic Variables.. 37
Dependent Variables: Behavioural Intentions (BI) and Broadband
     Adoption Behaviour (BAB)... 41
Usage of Broadband 42
Impact of Broadband.. 42
Summary 44
References 45

Section 1.2: The Methodological Underpinning

Chapter 3
Research Methodology **51**
Underlying Epistemology 52
Research Approaches 54
Survey Research Approach 58
Data Analysis 69
Summary... 71
References 72

Chapter 4
Development of Survey Instrument: Exploratory Survey and
Content Validity **76**
Conceptual Model 77
Instrument Development Process 78
Stage 1: The Exploratory Survey 79
Stage 2: Content Validation... 84
Stage 3: Instrument Testing 90
Final Survey Instrument 98
Summary... 100
References 101

Chapter 5
Development of Survey Instrument: Confirmatory Survey............**117**
Response Rate and Non-Response Bias 118

Reliability Test 119
Factor Analysis 120
Test for Ordering of Questionnaire Items 125
Discussions 126
Summary 131
References 132

Section 1.3: The Empirical Underpinning

Chapter 6
Empirical Findings: Adoption, Usage, and Impact of Broadband 135

Respondents' Profile 136
Adoption of Broadband 138
Usage of Broadband 153
Impact of Broadband 160
Summary 162
References 163

Chapter 7
Comparing the Current and Future Use of Electronic Services 164

Adoption and Use of Electronic Services and Applications 165
Conclusion 171
References 171

Chapter 8
Reflecting Upon the Empirical Findings: Validating the
Conceptual Model 172

Research Hypotheses 173
Broadband Adoption 173
Usage of Broadband 186
Impact of Broadband 189
Summary 192
References 192

Section 1.4: The Emerging Issues

Chapter 9
Exploring the Role of Broadband Adoption and Socio-Economic
Characteristics in the Diffusion of Emerging
E-Government Services 197

Background 198
Findings 202
Conclusion 206
References 207

Chapter 10
Broadband Quality Regulation: Perspectives from UK Users **209**
  Elizabeth A. Enabulele, Brunel University, UK
  Gheorghita Ghinea, Brunel University, UK
Introduction 210
An Integrated Broadband Quality Framework 211
Research Methodology 214
Research Findings 215
Concluding Discussion 219
References 220

Chapter 11
Conclusion: Contributions, Limitations, and Future
Research Directions **222**
Research Overview 223
Main Conclusions 227
Research Offerings and Implications... 229
Research Limitations 235
Future Research Directions.. 237
Summary 238
References 238

Division 2: The Dutch Case Study

Chapter 12
A Longitudinal Study to Investigate Consumer/User Adoption and
Use of Broadband in the Netherlands **241**
  Karianne Vermaas, Dialogic/Utrecht University, The Netherlands
  Lidwien van de Wijngaert, Utrecht University, The Netherlands
Introduction... 242
A Basic Model of Technology Adoption and Use 243
Research Methods.. 246
Results 247
Discussion and Conclusion 254
References 258

Chapter 13
Broadband in Dutch Education: Current Use, Experiences, and
Thresholds **261**
  Karianne Vermaas, Dialogic/Utrecht University, The Netherlands
  Sven Maltha, Dialogic Innovation & Interaction, The Netherlands
Introduction 262
Current Subscriptions to Broadband and Applications 263

Theoretical Framework 267
Research Goal, Questions, and Methods 270
Results 272
Summary 278
Discussion and Conclusion 278
Acknowledgment 281
References 281

Division 3: Developing Country Perspectives and Implications

Chapter 14
Factors Affecting Consumer Adoption of Broadband in Developing
Countries **285**
Theoretical Basis 286
Research Methodology 287
Findings: Broadband Adoption in Bangladesh 288
Findings: Broadband Adoption in the KSA 290
Conclusion 293
Acknowledgment 294
References 294

Chapter 15
Implications and Future Trends **296**
Implications for the Government 297
Implications for ISPs/Broadband Providers 298
Implications for the Content Providers 300
Future Trends 301
References 303

About the Contributors **304**

Index **307**

Foreword

From the perspective of an industry player, the first commercial priority is to understand my customers so that I can deliver what they want at a profit. That is no easy task and many ISPs are too small to afford the research that achieving such understanding requires. So, this book provides a welcomed leg-up for those who have the perseverance to digest this material.

I spent nearly 20 years as an economist and pricing manager at Telstra, the incumbent telecommunications carrier in Australia. There, I saw how effective market research can be in predicting customer behaviour. The current management team under CEO, Sol Trujillo, obviously thinks so too, employing large-scale market research to redefine its customer segments. As the broadband market matures and industry growth slows, it will be those who best understand their customers that will prosper most.

This book is not a handbook for our industry on how to conduct market research. Within the limitations of university based research, only the survey instrument could be used. With the greater resources available to leading industry operators, other techniques could also be used, which can only be mentioned in this book. But, the way in which the survey instrument has been used is carefully explained and there is an extensive review of previous related research.

For industry practitioners, whose time is very limited, Chapter 11 is a good place to start. It summarises the preceding chapters and helps draw out some of the implications of these finding for industry. For example, there are some hints about using such research to segment the market to provide better entry-level pricing to poorer customers. But, developing this idea is beyond the scope of this study.

I think industry readers would probably find Chapters 6 to 11 in the empirical section to be the most rewarding. Some of it will confirm what you already know or suspect, such as the importance of age, education, and income to the adoption of broadband. But, here is more evidence based on new and original survey work.

The hypotheses about what drives customer adoption and use of broadband are important for governments as well as industry players. This research is still in its infancy and much of what is being done is commercially sensitive and not in the public domain. So, the publication of this book is a very welcomed contribution to understanding our customers.

John de Ridder
Telecommunications Economist
www.deridder.com.au

Preface

Governments all around the world are encouraging broadband internet connectivity amongst its citizens. This is because broadband is portrayed as a means to increase the international competitiveness of a country and its role is seen as being critical to the growth and adoption of electronic commerce and other advanced e-services and applications, such as VoIP. Broadband connectivity can benefit a national economy in a number of ways as it has the potential to deliver economic value, public value, efficiencies in the public sector and improving citizens' lives. The prospective benefits of broadband clearly suggest that accessibility, adoption and use of broadband are likely to transform and affect almost every aspect of everyday life. Therefore, in order to realise the full potential that broadband offers, a number of countries have considered/are considering the diffusion and use of broadband technologies as an important item in their policy agendas.

Nevertheless, despite a large investment in developing the infrastructure and the provision of broadband access at affordable prices, the demand for broadband has not increased as expected in many countries around the globe. The slow rate of broadband adoption can be considered to be demand constrained in the developed countries where high speed access is made available to the majority of the population. Broadband penetration differs greatly across countries; for example, according to recent Organisation of Economic Cooperation and Development (OECD) statistics, some countries such as South Korea have reached a steady state, whilst others such as Greece represent much lower levels of penetration (OECD, 2005). This means that in order to enhance the homogenous adoption and use of broadband and to reduce the digital divide, it is essential to focus upon understanding the consumer or micro level factors influencing the deployment and use of broadband.[1]

The aim of this book is to provide an understanding of consumer-level factors affecting the adoption of broadband. It also aims to understand the usage of broadband and its impact upon consumers. The book is expected to contribute towards theory, practice and policy. The theoretical contribution of this research is that it synthesises the existing literature in order to enhance knowledge of broadband adoption, usage and impact from a global perspective. The theories that the book has specifically contributed to include diffusion of innovations, technology acceptance theories such as the technology acceptance model (TAM) and the theory of planned behaviour (TPB), policy making for telecommunications and consumer behaviour. Considering the slow adoption of broadband, it can be learnt that the policy makers and the providers of the innovation—in this case the telecommunications industry—hold a particular interest in the findings of this study. Policy makers are currently investigating how to increase the diffusion of broadband within their own country, and so information on other countries' experiences can prove useful. The telecommunications industry is interested in determining how to improve their current strategies. Therefore, for both policy and practice, this book will offer an understanding of the broadband diffusion strategies at both the macro and micro levels.

This is particularly useful as there is little research published in the area of consumer adoption, usage and impact of broadband. Understanding the usage and impact of broadband will be helpful for content-developing organisations to integrate compelling content with new generation broadband and will also help broadband service providers to improve their service.

The book is likely to enhance the understanding of how to encourage the adoption of broadband and other emerging telecommunication technologies and applications and will also help to formulate a strategy to bridge the digital divide. Therefore, the book will be an interesting read for various audiences, including broadband service providers, policy makers, academics/researchers, taught and research students of information systems, IT/telecom and marketing management. New doctoral students generally face problems in selecting and justifying the theories, models and research approaches in order to undertake and accomplish his/her research. The content of this book illustrates a systematic review of the various theories, models and research approaches which led to the selection and justification of the appropriate constructs from the different theories/models and, eventually, led to the development of a conceptual model. This book also illustrates the systematic selection and justification of a relevant research approach to test the conceptual model. This suggests that this book will provide very useful reading for new researchers including doctoral candidates.

In order to cater for the information needs of diverse readers and also to effectively deal with this complex but emerging topic, the book is structured into four divisions comprising of 15 chapters. The first division is divided into four sections. A brief description of each division, section and chapter is provided below.

Division 1, entitled *The United Kingdom Case Study* examines the consumer adoption, usage and impact of broadband in UK households. This division also discusses the emerging issues such as the impact of broadband on awareness and adoption of new electronic services and also the consumer perception of the service quality of broadband subscription in the UK. This division is further organised into four sections as described below:

The **first section**, entitled *The Theoretical Underpinning*, establishes the research problem, looks at the relevance of the research, defines the research aims and objectives, reviews the relevant technology adoption theories and models, and finally develops a conceptual model of broadband adoption, usage and impact. Section I consists of the following two chapters: Chapter 1, *Introduction*, and Chapter 2, *Conceptual Model for Examining Consumer Broadband Adoption*. Chapter 1 is useful for all potential readers (i.e., broadband service providers, policy makers, academics/researchers and taught and research students). Chapter 2 is particularly useful for specific audiences, such as academics/researchers and taught and research students.

Chapter 1 provides an overview for the research area. The chapter first describes the research problem and outlines the scope of this study. It then analyses the state of the research in the area of broadband adoption, usage and impact from a household consumer perspective. This leads to an outline of the research aims and objectives that this research has addressed, followed by a brief description of the research methodology that was utilised to conduct this research. The chapter then outlines the contributions that this research will make.

Chapter 2 first reviews and assesses the appropriateness of previous technology adoption models and constructs used to study broadband diffusion. It then provides further theoretical justification for selecting the constructs that are used to study broadband diffusion formulates the research hypotheses and finally draws a conceptual model of broadband adoption. The chapter also discusses the usage and impact aspect of broadband diffusion and identifies constructs for empirical investigations.

The **second section**, *The Methodological Underpinning*, primarily focuses upon determining the appropriate research approach employed to undertake this research. This section also describes the development of the survey instrument employed for data collection. Section II consists of the following three chapters: Chapter 3 *Research Methodology*; Chapter 4 *Development of Survey Instrument: Exploratory Survey and Content Validity*; and Chapter 5 *Development of Survey Instrument*. Although section II provides useful readings for all potential readers, it is particularly useful for audiences such as marketing professionals from broadband service providers industry, academics, researchers and students.

Chapter 3 aims to discuss research approaches in general and those specific to this research. It also provides the justification for the chosen research methodology, as well as detailed discussions on the specific methodological approach employed.

Chapter 4 aims to describe the development of a research instrument that is designed to investigate broadband adoption, usage and impact within UK households. The chapter describes the following three stages of developing a reliable research instrument: (1) identification of the factors from the literature that are expected to explain the broadband adoption behaviour and determining them by employing an exploratory survey; (2) content validation on items that result from the exploratory survey, the purpose of this step being to confirm the representativeness of items to a particular construct domain; and (3) a description of a pre-test and a pilot-test in order to confirm the reliability of measures.

Chapter 5 presents findings obtained from a confirmatory survey that was conducted in order to examine the adoption, usage and impact of broadband in UK households. The chapter provides an illustration and discussion of the estimation of response rates, non-response bias, reliability, construct validity and the effect of ordering questionnaire items. This chapter first discusses the appropriateness of response rates and issues of the non-response bias of the survey in light of existing work. Then it discusses the instrument validation process by reflecting upon issues such as content validity, reliability and construct validity.

The third section, *The Empirical Underpinning*, provides the fruit of this research monograph. This section presents and illustrates the findings obtained from confirmatory survey data collection before reflecting upon these findings. Section III consists of the following three chapters: Chapter 6 *Empirical Findings*; Chapter 7 *Comparing the Current and Future Use of Electronic Services* and Chapter 8 *Reflecting Upon the Empirical Findings*. Section III is recommended as being essential and useful reading for all potential readers.

Chapter 6 presents the findings obtained from the survey that was conducted to examine the adoption, usage and impact of broadband in UK households. An illustration of these findings comprises of descriptive statistics, differences between the adopters and non-adopters of broadband, demographic differences and regression analysis. This chapter also presents findings related to the usage of broadband and its effects on consumers' time allocation patterns on various daily life activities.

Chapter 7 illustrates the adoption and use of 41 online services belonging to seven different categories which were included to examine the current and future use at home and in the work place in the UK. These seven categories comprised communications (five online services), information seeking (seven online services), information producing (four online services), downloading (six services), media streaming (five services), e-commerce (eight services), and other activities that included entertainment activities (four services), social and personal (two services) and e-government.

Chapter 8 discusses and reflects upon the findings obtained in Chapters 6 and 7 from the theoretical perspectives presented in Chapter 2. It also discusses the empirical issues that have been reported from the survey findings. This chapter first discusses the refined and validated model of broadband adoption. Finally, the chapter provides a discussion on the usage of broadband and how it affects consumers' time allocation patterns on various daily life activities.

Finally, the **fourth section**, *The Emerging Issues*, explores some upcoming issues critical to the area of broadband adoption, usage and impact and also concludes the UK case study. Chapter 9 explores the role of broadband adoption and socio-economic characteristics in the diffusion of emerging e-Government services. Elizabeth Enabulele and Gheorghita Ghinea (Brunel University, UK) present a critical topic entitled *Broadband Quality Regulation* (Chapter 10). Finally, Chapter 11 *Conclusions*, formulates the conclusions of the UK case study and provides a discussion on the contributions of this research, as well as its limitations and future research directions. Section IV is also recommended as being essential and useful reading for all potential readers.

Chapter 9 examines citizens' awareness and adoption of e-Government initiatives, specifically the 'Government Gateway' in the United Kingdom. Since these services have been recently introduced, an investigation was needed to study if the demographic characteristics and home internet access are affecting the awareness and adoption of these services. Therefore, the second aim of this chapter was to examine the effect of citizens' demographic characteristics and home broadband access on the awareness and adoption of e-Government services. To fulfil the specified aims, this study undertook an empirical examination of the awareness and adoption of the 'Government Gateway' amongst UK citizens.

Finally, **Chapter 10** emphasises the fact that in the rush to achieve market share, insufficient attention has been paid to quality issues, which forms the central theme of the chapter. Indeed the concept of quality is a multi-faceted one, for which various perspectives can be distinguished. The chapter explores broadband quality as perceived by users in the UK and reports the results of a survey, which determined the users' perception on broadband quality. The results of the survey show that quality, although desired by many, has been neglected in favor of the desire to have access to the Internet via broadband at the lowest cost possible. However, this has not encouraged some consumers to take on broadband access despite some low prices offered by service providers, as these low prices for broadband access is not commensurate to their needs.

Chapter 11 summarises the research findings and provides a discussion of the research contributions and implications of this research in terms of the theory, policy and practice. This chapter also delineates the research limitations and presents future research directions in the area of broadband diffusion and adoption.

Division 2, entitled *The Dutch Case Study,* examines the consumer adoption, usage and impact of broadband in the Netherlands. Division II consists of two chapters. Karianne Vermaas and Lidwien van de Wijngaert (Utrecht University in the Netherlands) present research outcomes from a longitudinal study on consumer/user adoption and usage of broadband in the Netherlands (Chapter 12). In Chapter 13, Karianne Vermaas and Sven Maltha (Dialogic Innovation and Interaction in the Netherlands) report on current usage, experiences and thresholds of Broadband in Dutch education. Division II is also recommended as essential and useful reading for all potential readers. A more detailed description of Chapters 12 and 13 is provided below:

Chapter 12 investigates the central research question: How do Dutch internet users with a broadband connection differ from people with a narrowband connection in terms of demographics (age, gender, education), internet experience (experience, frequency, intensity of use), expectations (of narrowband users), experiences (of broadband users), annoyances and patterns of internet usage? This chapter also addresses the question of whether and how these differences change over time. The chapter uses a model of technology adoption and use that is built upon different theories such as diffusion of innovations, uses and gratifications, and media choice theory.

Chapter 13 explores the impact of broadband on the following aspects of education from a user perspective: (1) to what extent broadband is used in Dutch education (in the class room as well as in the organisation as a whole) and (2) the experiences that teachers have with broadband, including impediments and added value. This issue was investigated employing a survey data collection from 221 Dutch teachers, ICT-coordinators and school boards. The results show that teachers, ICT coordinators and school boards are interested in using broadband in their schools as they see the added value, but there seems to be an impasse: without infrastructure, there are no services and without services there is no need for infrastructure. Schools can break out of the causality dilemma by giving an impulse to the market by combining forces and demand. Moreover, teachers need to be trained in using the new tools and service.

Division 3, entitled *Developing Country Perspectives and Implications,* examines consumer adoption of broadband in the developing countries of Bangladesh and the Kingdom of Saudi Arabia (KSA). This division also discusses the further implications of the findings and future trends in the area of adoption and diffusion of broadband. Division III provides essential readings for all potential readers, particularly for audiences from developing countries such as marketing professionals from broadband service providers industry, academics, researchers and students. This division is further organised into two chapters as described below:

Chapter 14 examines the factors affecting the adoption of broadband in two developing countries, namely, Bangladesh and the Kingdom of the Saudi Arabia. This chapter undertook an empirical examination using a survey approach for examining the adoption of broadband.

Finally, **Chapter 15** presents implications of the research presented in this book and outlines the future research trends in the area of adoption and diffusion of broadband. The findings of the research detailed in this book generate a number of implications that may be relevant to policy makers, ISPs and other relevant stakeholders for increasing consumer adoption of broadband.

This book provides exhaustive coverage on a particular research issue and the author hopes that this will provide a positive contribution to the area of information systems in general and, specifically, to the adoption and diffusion of broadband amongst consumers.

However, in order to make further research progress and improvement in the area of adoption and diffusion of broadband and other consumer technologies, I would like to welcome feedback and comments about this book from readers. Comments and constructive suggestions can be sent to me care of Idea Group Inc. at the ad-

dress provided at the beginning of the book, or by electronic mail to: ykdwivedi@
gmail.com.

Yogesh Kumar Dwivedi
Swansea, UK
May 2007

Endnote

[1]   Organisation for Economic Cooperation and Development, OECD Broadband
    Statistics, June 2005, http://www.oecd.org/document/16/0,2340,en_2649_
    34225_35526608_1_1_1_1,00.html#data2004

Acknowledgment

Guru Brahma Gurur Vishnu, Guru Devo Maheshwaraha
Guru Saakshat Para Brahma, Tasmai Sree Gurave Namaha

Meaning: Guru (the teacher) is verily the representative of Brahma (the creator), Vishnu (the sustainer) and Shiva (the destroyer). He creates, sustains knowledge, and destroys the weeds of ignorance. I salute such a Guru.

The ideas presented in this book have been developed as a result of advice and assistance offered by a number of people. It is by the grace of God and the will of the Almighty that I take this opportunity to convey my regards and thanks to those who have helped me and supported me at different stages in the completion of the book.

The first draft of this manuscript was written whilst I was a PhD student in the department of information systems and computing (DISC) at Brunel University. I would like to thank a number of academic staff at DISC, particularly Dr. Jyoti Choudrie, Professor Guy Fitzgerald, Professor Ray Paul, and Dr. Willem-Paul Brinkman for their positive criticism and suggestions during the different stages of the research presented in this book. Furthermore, I wholeheartedly acknowledge the continuous inspiration and encouragement rendered to me by my friends and colleagues at DISC.

The participants in this research set aside time in order to complete the question-naire and to provide insightful comments; I am highly grateful to them all for their invaluable assistance.

This book would not have been possible without the cooperation and assistance of the staff at IGI Global. I would like to thank my associates at IGI Global, namely Michelle Potter for handling the book proposal, Jan Travers for managing the contract, Meg Stocking, and Lynley Lapp for managing this project and especially for answering queries and helping to keep this project on schedule. A special word of thanks also goes to Ms. Kristin Roth and the anonymous reviewers for their useful and construc-tive comments that have been incorporated into the final version of this book.

Last, but not least, I bestow my unbounded gratitude and deepest sense of respect to my family whose blessings, concerted efforts, constant encouragement, and wholehearted co-operation enabled me to reach this milestone.

Divison I

The United Kingdom
Case Study

Section 1.1

The Theoretical Underpinning

Chapter 1

Introduction to Broadband Adoption and Usage Research

Abstract

This chapter provides an introduction to this research monograph. The following section provides a definition of broadband, outlines broadband's potential, and then provides a discussion of the research problem comprising the focus of this book. Following that, a discussion of the state of broadband adoption, usage, and impact is presented. This is ensued by the definition of the research aim and objectives. Details are provided concerning the research approach that will be utilised in order to achieve the proposed aims and objectives. The research contributions of this study are then offered. Finally, a summary of this chapter is provided.

Defining the Research Problem

Defining Broadband

A term frequented within this book is 'broadband,' which indicates that a short explanation is warranted at the outset. The term 'broadband' itself has no established definition. It varies amongst countries (Firth & Kelly, 2001) and evolves over time, as the underlying transmission and routing technologies continuously advance; yesterday's broadband is considered to be today's 'narrowband' (Sawyer et al., 2003). Instead of referring to a single technology, the umbrella term of 'broadband technology' embraces a variety of high-speed access technologies, including asymmetric digital subscriber line (ADSL), cable modems, satellite, and wireless fixed (Wi-Fi) networks (Sawyer et al., 2003).

Given the variations in defining 'broadband,' for the purpose of this research we follow the technology neutral definition suggested by the Broadband Stakeholder Group (BSG) that defines broadband as:

always on access, at work, at home or on the move provided by a range of fixed line, wireless and satellite technologies to progressively higher bandwidths capable of supporting genuinely new and innovative interactive content, applications and services and the delivery of enhanced public services. (BSG, 2001)

This definition was chosen as it is technology neutral. That is, it has less to do with the technical speed and instead focuses on functionality, which has more to do with what a user can do with broadband (Sawyer et al., 2003). According to Ofcom (2005), the term 'broadband' refers to higher bandwidths and 'always on' services offering data rates of 128 kbps and above. Dial-up or narrowband refers to Internet access that offers speed equal to or below 128 kbps (Ofcom, 2005).

Outlining the Potential of Broadband

Since the emergence of the Internet, broadband is being considered as the most significant evolutionary step. It is considered to be a technology that will offer end users with fast and 'always on' access to new services, applications, and content with real lifestyle and productivity benefits (Sawyer et al., 2003). International organisations, such as the International Telecommunication Union (ITU) and the Organisation for Economic Co-operation and Development (OECD) forecasted broadband to be a vital means of enhancing competitiveness in an economy and also for sustaining economic growth (BSG, 2004; ITU, 2001; OECD, 2001; Oh et al., 2003). According to a report from the United Kingdom Broadband Stakeholder Group, broadband provides a number of ways of enhancing a nation's economy and the quality of a citizen's life, as it stated that:

...Full exploitation of broadband-enabled ICT, content, applications and services can help the UK to become a truly competitive knowledge-based economy and can be leveraged to help the UK's citizens become healthier, better educated and more engaged in their communities and society. ...Societies that adopt, adapt, and absorb the benefits of broadband enabled ICT, services and applications quickly and deeply will achieve significant benefits in terms of productivity, innovation, growth and quality of life as well as significant competitive advantage over societies that don't... (BSG, 2004)

The focus group BSG had the forethought that broadband would benefit a national economy in the following four ways: by delivering economic value, delivering public value, delivering efficiencies in the public sector, and improving people's lives (BSG, 2004). Examples of delivering economic values include the potential of improving the productivity and competitiveness of SMEs by employing broadband as a communication channel for capturing emerging business opportunities (BSG, 2004). Broadband also offers potential benefits to larger companies, as it provides an efficient supply chain and economical ways to do business with customers, such as online retailing (BSG, 2004). Furthermore, since broadband facilitates home-working, it can also save costs in office space for both large and small organisations (BSG, 2004).

Similar to commercial organisations, broadband also offers the potential to governments in creating electronic services and delivering them to citizens in a cost effective and transparent manner. By exploiting the potential of broadband, direct benefits can be obtained that include improvements in education delivery, empowering patients, effective utilisation of patient records, and employing cost effective broadband for enabling tele-monitoring health applications in order to reduce the number of patients in hospitals (BSG, 2004). Such electronic services in the public sector will specifically help to reduce the cost of delivery, increase the quality of healthcare, thereby increasing the citizens' trust and confidence in public services and generally in the government (BSG, 2004). The 2004 Gershon spending review suggested a direct link between information and communications technology (ICT) investment made by the government and the ability of ICTs in delivering efficient savings in the government (BSG, 2004).

Broadband can also potentially improve citizens' lives in several ways. It can help equip children with ICT skills for employment purposes and improve the way they obtain education. Similarly, since broadband facilitates working at home (homeworking), it can help people to obtain a better work/life balance that is characterised by more empowerment, more productivity and less stress. Broadband also offers direct benefits to elderly people, as it can be utilised to provide personalised care at home, hence, removing the need to live in hospitals or care homes (BSG, 2004).

The discussion on the potential of broadband clearly suggests that accessibility, adoption, and use of broadband are likely to transform and affect almost every aspect of a citizen's everyday life (Oh et al., 2003). Therefore, in order to harvest its full potential, it is appropriate to understand the deployment and adoption of such an emerging technology.

Outlining the Research Problem

Although broadband offers several advantages to the public and private sectors in terms of cost savings, efficiency, and competitiveness, the shift to broadband requires massive investments in terms of new networks and infrastructures, along with the development of new content, services, applications, and business models. As discussed, since broadband diffusion is regarded as a measure of international competitiveness (BSG, 2004; Langdale, 1997; Oh et al., 2003; Sawyer et al., 2003), many governments around the world have set ambitious targets for the deployment of broadband services (BAG, 2003; National Broadband Task Force, 2001; Office of the e-Envoy, 2001; Office of Technology Policy, 2002). This is because the high penetration rate of broadband is perceived to have a positive impact on the growth and development of the Internet, electronic commerce (e-commerce), and the information economy (Lee et al., 2003; Sawyer et al., 2003). The United Kingdom (UK) government believes that the rapid rollout and adoption of broadband across the nation is important to both its social and economic objectives (Oftel, 2003); hence it has made a commitment to making the UK the most competitive and extensive broadband market in the G7 (Office of the e-Envoy, 2001). Therefore, governments of a number of countries including South Korea, Japan, Hong Kong, Sweden, Canada, and also the UK have made large investments for developing a broadband infrastructure that will deliver high speed Internet access to end users, including household consumers and SMEs (BSG, 2004; OECD, 2001; Oh et al., 2003; Sawyer et al., 2003).

Despite a large investment for developing the infrastructure, the majority of countries have had slow adoption rates of broadband (OECD, 2001; Oftel, 2003). Although the UK initiated an early rollout of infrastructure competition, the rate of broadband adoption has been relatively slow since the start (OECD, 2001). According to an Oftel (2003) report, until the year 2003, the growth rate of broadband was slower than other similar European and North American countries such as Sweden, France, Germany, and the USA (Oftel, 2003). Furthermore, Oftel's (the UK's communication regulator until 2003) international benchmark study (Oftel, 2003) also suggested that the percentage growth of residential broadband subscriptions was slowing down in all countries including the UK. Due to intense competition amongst Internet service providers (ISPs), it has been found that there was a sharp decrease in price (i.e., a monthly subscription fee offered to residential consumers) between 2002 and 2003 (Oftel, 2003). It has been suggested that when the numbers of cable modem connections are excluded, in terms of the price of broadband, the UK is similar to France and cheaper than all other similar European and North American countries (Ofcom, 2004). This sharp decrease in price led to sudden increases in household broadband subscriber numbers to 3.2 million (12 percent of households) (Ofcom, 2004).

In the summer of 2005, the UK's incumbent monopolist British Telecommunication (BT) reached 99.6 percent in terms of broadband coverage of the households in the UK; however, the total number of reported broadband subscribers have reached only 3.99 million (15 percent percent of households) (Ofcom, 2005). This is only a 3 percent increase since last year, which suggests a further slow down in the adoption rate of broadband. A recent BBC News article has found that the market research organisation Point Topic has suggested that there is a "considerable slow-down in the rate at which households adopt the Internet" (BBC News, 2005). According to Point Topic's study, existing Internet users are upgrading their connections to get more out of them. However, new users are "unlikely to sign up a fast connection straight away" (BBC News, 2005). Such saturation is already visible in South Korea and Hong Kong (BBC News, 2005). The few recent reports and news have also indicated heterogeneous adoptions, possible digital divides and digital choices in terms of broadband adoption (BBC News, 2005; Dutton et al., 2005; Kinnes, 2005). These studies suggest that there are socio-economic factors that are affecting the homogenous adoption of broadband.

Despite the provision of broadband access at affordable prices, the demand for broadband has not increased as expected in many countries around the globe. Researchers are suggesting that the provision of broadband is more 'demand constrained' than 'supply constrained' (Haring et al., 2002). This means that in order to enhance the homogenous adoption and use of broadband and to reduce the digital divide, it is appropriate to focus on understanding the factors influencing the decisions of household consumers (Crabtree, 2003; Oh et al., 2003; Stanton, 2004). Previous research undertaken on the adoption of technology, such as the adoption of Personal Computers (PCs) by residential consumers, has also emphasised the role of the demand perspective (Venkatesh & Brown, 2001). The discussions presented have provided the motivations for conducting research on the consumer adoption of broadband, its usage and impact in UK households. This was essential because before formulating the aims and objectives of the following section, an understanding of the existing research in the emerging area of broadband adoption, usage and impact is necessary.

The State of Broadband Adoption, Usage, and Impact Research

Adoption Studies

The adoption studies discussed henceforth provide mainly discussions of the macro factors that drive the success or slow uptake of broadband. As stated earlier, research on the topic of broadband adoption at the micro level is minimal (Crabtree, 2003; Oh et al., 2003; Stanton, 2004). In an initial study of broadband deployment in South Korea, Lee et al. (2003) identified three major factors comprising public sector actions, private sector actions, and the socio-cultural environment factors that explained the high rate of broadband adoption in South Korea. Further research suggested that six success factors are responsible for driving the high penetration rate of broadband within the South Korean residential consumers (Choudrie & Lee, 2004; Lee & Choudrie, 2002). These six key factors consist of the government's vision, strategy and commitment, facilities-based competition, pricing, the PC Bang phenomenon, culture and geography and demographics (Choudrie & Lee, 2004; Choudrie et al., 2003a,2003b). To obtain a more balanced view, the UK perspective was also investigated. Dwivedi et al. (2003a,2003b) examined the ISPs views on factors affecting broadband adoption in the UK. This exploratory study suggested that a high price, lack of content and awareness are the factors that are severely affecting the adoption of broadband amongst the UK residential consumers (Dwivedi et al., 2003a, 2003b).

However, now adoption studies on consumers have begun to emerge. Amongst the initial studies is one by Oh et al., (2003). This study examined the individual level factors affecting the adoption of broadband access in South Korea by combining factors taken from Rogers' diffusion theory and the technology acceptance model (Oh et al., 2003). The findings of this study suggest that congruent experiences and opportunities in adopting a new technology affect users' attitudes through the three extended technology acceptance model constructs, namely perceived usefulness, perceived ease of use, and perceived resources (Oh et al., 2003). Stanton (2004) has also analysed the secondary data of the USA consumers in order to study the digital divide and suggested an urgent need to understand the demography and other factors of broadband adopters and non-adopters in order to increase the growth rate of broadband and also to bridge the digital divide (Stanton, 2004).

Usage and Impact Studies

Usage is the other topic of interest in the information systems (IS) area and is pertinent to this research. Studies in this area have been in the form of user surveys that have examined broadband users' behaviour in comparison to that of narrowband users (Anderson et al., 2002; Bouvard & Kurtzman, 2001; Carriere et al., 2000; Dwivedi & Choudrie, 2003a, 2003b; Horrigan et al., 2001). Results from these surveys suggest that Internet users behave differently when they have broadband access (Carriere et al., 2000; Dwivedi & Choudrie, 2003a). Broadband users use the online facilities on a longer basis, utilise more services or applications and apply them more often (Carriere et al., 2000; Dwivedi & Choudrie, 2003a; Horrigan et al., 2001). The majority of broadband users rate their online experience as compelling (Anderson et al., 2002; Carriere et al., 2000; Dwivedi & Choudrie, 2003a; Horrigan et al., 2001). In comparison to dial up users, broadband users spend more time, in total, on electronic media applications, such as online music (Bouvard & Kurtzman, 2001). Surveys conducted on broadband users also suggest that these users make more online purchases and procure more varied categories of products in comparison to narrowband users (Carriere et al., 2000; Dwivedi & Choudrie, 2003b). Although these studies examine the usage of broadband, they lack theoretical underpinnings as they are led by data and are exploratory in nature (Oh et al., 2003). Understanding the impact of broadband usage on consumers' daily life is still untouched by previous studies.

The discussion on broadband adoption and usage suggests that there were few efforts made to examine broadband diffusion from the household consumer perspective. Furthermore, the aforementioned studies were not statistically tested to determine the differences between broadband consumers and narrowband consumers. As a result, these studies lack statistical conclusion validity in their findings (Straub et al., 2004). Another observation from a review of previous literature suggests that existing studies examined exploratory issues related to either the adoption or usage of broadband. None of them, however, provide a thorough understanding of all three components of broadband diffusion (i.e., adoption, usage, and impact) from the household consumer perspective. Therefore, both discussions on the aforesaid research problem and the lack of studies on broadband adoption, usage, and impact provide the motivation for this study. Before outlining the aims and objectives in the following section, the following subsection defines the scope of this study.

Scope and Definitions

Whilst studying the diffusion of broadband, there are many stakeholders to consider, including the government, ISPs, business consumers, public organisations, and residential consumers (Choudrie et al., 2003a, 2003b). The previous discussions clearly suggest broadband adoption is demand-constrained and existing studies have not yet addressed the issue of consumer adoption, usage, and the impact of broadband in UK households. The focus of this study is therefore the UK household consumer of broadband. Thus, the proposed conceptual model (Chapter 2) will only consider factors and studies that are relevant to household consumers. As previously mentioned, a focus upon the household consumers is considered to be imperative for this study and the two main reasons for this are as follows. First, as existing studies have mainly considered only the supply-side stakeholders, there has been little attention paid to the examination of consumers. Second, at this stage of broadband implementation, the supply-side factors are not considered to be a problem, however, growth is constrained by the demand side as consumers are reluctant to subscribe to the technology in question.

Since the focus of this research is the consumers, at this point the differentiating factor between the terms consumers and users is provided. According to Rice (1997), 'consumers' are those who pay for services and goods, whilst 'users' are individuals who are affected by or who affect the products or services. In other words, users are those who use the products and services but do not pay for them (Rice, 1997). For example, a child can be categorised as a user since he/she uses broadband for online gaming and to undertake homework, however, the child does not pay for the service. In contrast, the parent can be identified as a consumer as well as a user since he/she pays and uses the service.

For the purpose of this research, the term 'diffusion' is defined as "the process by which an innovation is communicated through certain channels over time among the members of a social system" (Rogers, 1995, p. 5). The term 'innovation' refers to "an idea, practice or object that is perceived as new by an individual or other unit of adoption" (Rogers, 1995, p. 11). The meaning of 'newness of an idea' or 'innovation' is likely to differ from consumer to consumer. If the idea seems new to an individual, it is an innovation (Rogers, 1995). For example, broadband may not be an innovation for someone who has already subscribed to it. However, if someone does not know about broadband or has not subscribed to it, then, for such a consumer, broadband would be considered as an innovation. Since a large number of consumers in the UK and other countries have not yet subscribed to broadband, for the purpose of this research it is considered as an innovation. The study of diffusion involves three components: the adoption of a new innovation, its usage and the subsequent impact of its usage (Rogers, 1995). Therefore, this study will examine the constructs relevant to broadband adoption, usage, and its impact on household consumers in the UK.

Research Aims and Objectives

Researchers in the IS field have been studying the adoption of ICTs at the organisational level (Venkatesh & Brown, 2001). However, until recently, IS research has less frequently focused upon the household adoption and impact of ICT. Due to the advent of the Internet within daily life, ICTs have become an essential part of everyday life (Oh et al., 2003). This has led researchers to conduct studies within the household context or, in other words, moving beyond the organisational level. Mingers and Stowell (1997) argue "information systems is much more than simply the development of computer-based business systems- electronic and information technology is now so fundamental within society that IS as a discipline must concern itself with the general evolution of human communication."

In line with this, one of the first studies within the IS field to examine the adoption of PCs in the household was recently undertaken (Venkatesh & Brown, 2001). To continue with this line of research, but within a different context, Anckar (2003) offered an understanding of the drivers and inhibitors to e-commerce adoption within Finnish households. Although such studies are becoming prevalent, they have not yet been extended to examine the adoption of emerging ICTs such as broadband. This is due to the fact that the technology in question — broadband — is still emerging in the market. It can be found that the majority of the research associated with the topic of broadband is still exploratory in nature, mainly focusing on the usage of the technology, and provides little insight into consumer adoption or rejection determinants. Sawyer et al. (2003) argued that "the differential rates of use and growth, suggest that understanding broadband connectivity is a complex milieu." Consequently, investigating the adoption, usage, and impact of broadband in the context of consumers is expected to make an incremental contribution to information systems research.

Bearing the aforementioned discussion in mind, the primary aim of this book is to: *investigate individual-level factors affecting consumers' adoption of broadband.* This study also aims to: *understand the usage of broadband and its impact upon consumers.* To achieve these aims, the following research objectives shall be undertaken:

1. To develop a conceptual model for examining consumer adoption, usage, and impact of broadband. This will be achieved, first, by reviewing the theories and models that focus upon individual and/or consumer adoption, usage, and impact of technology. The next step is to select relevant constructs from appropriate theories and models and to formulate a research hypothesis in order to examine broadband adoption, usage and impact from the UK household consumer perspective.

2. To operationalise the constructs included in the conceptual model by developing a research instrument and demonstrating their reliability and validity. The research instrument was developed and validated utilising content validity, reliability, and construct validity approaches.

3. To empirically validate and refine the proposed conceptual model in order to examine broadband adoption, usage and impact in households. This was achieved by collecting and analysing data from the household consumers.

4.   To provide implications for practice and policy that may encourage consumer adoption and use of broadband. Furthermore, there will be a reflection of the impact of consumers' use of broadband.

5.   To explore emerging issues such as perceived service quality, impact of broadband, longitudinal investigation of broadband usage, and adoption of broadband in developing countries.

Based on the aims and objectives, Chapter 2 delineates the research questions and hypotheses in order to understand the factors influencing consumer adoption of broadband, its usage, and impact in the context of individual consumers.

Research Approach

Since the research object is the consumer, it can be argued that the survey approach is the most suitable research approach for this study. This is due to issues such as convenience, cost, time, and accessibility (Gilbert, 2001). The extent to which a researcher can be a part of the context being studied is an important factor in determining the research approach.

Within the household context, it is difficult for a researcher to be a part of the context; therefore the survey approach was more feasible than others, such as case study, ethnography, and observations. Selection of the approach in this case is also influenced by the type of theory and models employed to examine research related to broadband adoption and diffusion (Chapter 2). A conceptual model proposed in Chapter 2 includes a number of research hypotheses that need to be tested before concluding this study. This requires collecting quantitative data and statistical analysis in order to test the research hypotheses.

Although there are a number of research approaches available within the category of quantitative positivist research (Straub et al., 2005), the survey is the only appropriate research approach that can be employed to conduct such research (i.e., that which requires hypotheses testing and validation of a conceptual model) in a social setting which, in this instance, is the household.

Furthermore, the aim of this research was to examine broadband adoption and diffusion across the UK, which was a nationwide perspective. Thus, in order to get an overall picture of the research issue, the collection of data from large numbers of participants from across the UK was essential. It means that employing any other approach, such as ethnography that utilises an interview or observation as data collection tools, demands huge amounts of financial resources, manpower, and time (Cornford & Smithson, 1996). All three factors were limited in this research project, and this restricted the ability of the researcher in employing them for this investigation. One of the planned contributions of this study is to provide insights to ISPs about the factors that are salient to consumer adoption and non-adoption of broadband, and to establish relationships between factors such as behavioural intention and actual adoption. In order to achieve this, it is vital to collect quantitative data on a number of variables including demographics and to perform a regression analysis to illustrate relationships between factors. This is another logical reason for adopting the quantitative approach via a survey and collating data that may assist ISPs in understanding the behaviour of household consumers and their demographic characteristics in order to encourage and promote broadband adoption.

The quantitative data was collected from UK household consumers. In order to collect empirical data from the target population, a self-administered questionnaire via postal mail was considered to be the most appropriate data collection method. The reasons for using the self administered questionnaire are: it addresses the issue of reliability of information by reducing and eliminating differences in the way in which the questions are asked (Cornford & Smithson, 1996); it involves relatively low costs of administration; it can be accomplished with minimal facilities; it provides access to widely dispersed samples; respondents have time to provide thoughtful answers; it assists with asking long questions or complex response categories; it allows asking of similar repeated questions; and, also, the respondents do not have to share answers with interviewers (Fowler, 2002).

The collated data was analysed using the statistical software SPSS version 11.5. In order to collate the data, it was appropriate to employ the UK-Info Disk V11 database as a sampling frame, stratified random sampling as a basis of sample selection, and a postal questionnaire as a data collection tool. Reasons for these selections are provided in Chapter 3. Issues relating to data analysis are also discussed in detail in Chapter 3, which suggests that a number of statistical techniques such as factor analysis, t-test, ANOVA, χ^2 test, discriminant analysis, linear and logistics regression analysis are appropriate for utilisation in data analyses.

Research Contributions

This research is expected to contribute towards theory, practice and policy. The theoretical contribution of this research is that it integrates the appropriate IS literature in order to enhance knowledge of technology adoption from the consumer perspective. It also contributes towards theory by empirically confirming the appropriateness of various constructs and validating a conceptual model in the context of household consumers. This study also contributes towards theory as it develops and validates a research instrument for data collection. The instrument examines broadband adoption, usage and impact. According to Straub et al (2004) it is essential to create and validate new measures or instruments in a situation where theory is advanced; however, prior instrumentation is not developed and validated. Such efforts are considered to be a major contribution to scientific practice in the IS field (Straub et al., 2004).

Considering the slow adoption of broadband, it can be learned that the policy makers and the providers of the innovation—in this case the telecommunications industry—hold a particular interest in the findings of this study. Policy makers are currently investigating how to increase the diffusion of broadband within their own country, and so information on other countries' experiences will be useful. The telecommunications industry is interested in determining how to improve their current strategies. Therefore, for both policy and practice, this study offers an understanding of the broadband diffusion strategies at a household consumer level. This is particularly useful as there is little research published in the area of consumer adoption, usage and impact of broadband.

Summary

This chapter provided an introduction to the research problem that this book encompasses, and established the scope of this research monograph. It discussed the existing research in the area of broadband adoption, usage and impact from a household consumer perspective, and went on to define the research aims and objectives that this research later addresses. A brief description of the research methodology and the contributions that this research will make were then presented.

A lack of studies on consumer level broadband adoption and diffusion has resulted in a lack of appropriate conceptual models specific to broadband. Therefore, the next chapter will develop a model to understand household diffusion by identifying appropriate constructs from previous theories and models on technology adoption. The chapter will also formulate several research hypotheses.

References

Anckar, B. (2003). Drivers and inhibitors to e-commerce adoption: exploring the rationality of consumer behaviour in the electronic marketplace. In C. Ciborra et al. (Eds.), *Proceedings of the 11th ECIS on New Paradigms in Organizations, Markets and Society*, Napoli, Italy.

Anderson, B., Gale, C., Jones, M. L. R., & McWilliam, A. (2002). Domesticating broadband-what consumers really do with flat rate, always-on and fast Internet access? *BT Technology Journal, 20*(1), 103-114.

BAG. (2003). *Australia's broadband connectivity: The Broadband Advisory Group's report to government*. Retrieved March 25, 2004, from http://www.dcita.gov. au/ie/publications/2003/01/bag_report

BBC News. (2005). *Broadband reveals digital divide*. Retrieved May 25, 2005, from http://news.bbc.co.uk/1/hi/technology/4483065.stm

Bouvard, P., & Kurtzman, W. (2001). *The broadband revolution: How super fast Internet access changes media habits in American households*. Retrieved October 25, 2002, from http://www.arbitron.com/downloads/broadband.pdf

BSG Briefing Paper. (2004). *The impact of broadband-enabled ICT, content, applications and services on the UK economy and society to 2010*. Retrieved November 15, 2004, from http://www.broadbanduk.org/news/news_pdfs/ Sept percent202004/BSG_Phase_2_BB_Impact_BackgroundPaper_ Sept04(1).pdf

BSG Report. (2001). Report and strategic recommendations. Retrieved November 30, 2002, from http://www.broadbanduk.org/reports/BSG_Report1.pdf

Carriere, R., Rose, J., Sirois, L., Turcotte, N., & Christian, Z. (2000). *Broadband changes everything*. Retrieved November 30, 2002, from *McKinsey & Company* at http://www.mckinsey.de/_downloads/knowmatters/telecommunications/broadband_changes.pdf

Choudrie, J., & Lee, H. (2004). Broadband development in South Korea: Institutional and cultural factor. *European Journal of Information Systems*, *13*(2), 103-114.

Choudrie, J., Papazafeiropoulou, A., & Lee, H. (2003a). A web of stakeholders and strategies: A case of broadband diffusion in South Korea. *Journal of Information Technology*, *18*(4), 280-303.

Choudrie, J., Papazafeiropoulou, A., & Lee, H. (2003b). Applying stakeholder theory to analyse the diffusion of broadband in South Korea: The importance of the government's role. In C. Ciborra et al. (Eds), *Proceedings of the 11ᵗʰ ECIS on New Paradigms in Organizations, Markets and Society*, Napoli, Italy.

Cornford, T., & Smithson, S. (1996). *Project research in information systems: A student's guide*. London: Macmillan Press Ltd.

Crabtree, J. (2003). *Fat pipes, connected people-rethinking broadband Britain*. Retrieved March 30, 2004, from iSOCIETY Report at http://www.theworkfoundation.com/pdf/1 843730146.pdf

Dutton, W. H., di Gennaro, C., & Hargrave, A. M. (2005). *The Internet in Britain: The Oxford Internet survey (OxIS)*, Retrieved July 20, 2005, from http://www. oii.ox.ac.uk/research/oxis/OxIS_2005_Internet_Survey.pdf

Dwivedi, Y. K. & Choudrie, J. (2003a). The impact of broadband on the consumer online habit and usage of Internet activities. In M. Levy et al. (Eds.), *Proceed-*

ings of the 8th UKAIS Annual Conference on Co-ordination and Co-opetition: The IS role. Warwick, UK.

Dwivedi, Y. K. & Choudrie, J. (2003b). Considering the impact of broadband upon the growth and development of B-2-C electronic commerce. In R. Cooper et al. (Eds.), *Proceedings of the ITS Asia- Australasian Regional Conference,* Perth, Australia.

Dwivedi, Y. K., Choudrie, J., & Gopal, U. (2003). Broadband stakeholders analysis: ISPs perspective. In R. Cooper et al. (Eds.), *Proceedings of the ITS Asia- Australasian Regional Conference,* Perth, Australia.

Firth, L., & Kelly, T. (2001). *Broadband briefing paper.* Retrieved July 25, 2003, from ITU at www.itu.int/broadband

Fowler, F. J., Jr. (2002). *Survey research methods.* London: Sage Publications.

Gilbert, N. (2001). *Researching Social Life.* London: Sage Publications.

Haring, J., Rohlfs, J., & Shooshan, H. (2002). *Propelling the broadband bandwagon.* Strategic Policy Research, Maryland.

Horrigan, J. B., & Rainie, L. (2002). *The broadband difference: How online Americans' behaviour changes with high-speed Internet connections at home.* Retrieved September 20, 2003, from http://www.pewInternet.org/pdfs/PIP_Broadband_Report.pdf

International Telecommunication Union. (2001). *A broadband future: Reconciling opportunities and uncertainties.* Retrieved November 20, 2002, from *ITU News,* 6, 3-7 at http://www.itu.int/itunews/issue/pdf/2001/06.pdf.

Kinnes, S. (2005). *One in three rejects technology.* Retrieved July 25, 2005, from *Times Online* at http://www.timesonline.co.uk/article/0,,2103-1687615,00.html

Langdale, J. V. (1997). International competitiveness in East Asia: Broadband telecommunications and interactive multimedia. *Telecommunications Policy, 21,* 235-249.

Lee, H., O'Keefe, B., & Yun, K. (2003). The growth of broadband and electronic commerce in South Korea: contributing factors. *The Information Society, 19,* 81-93.

Lee, H., & Choudrie, J. (2002). Investigating broadband technology deployment in South Korea. Brunel- DTI International Technology Services Mission to South Korea. DISC, Brunel University, Uxbridge, UK.

Mingers, J., & Stowell, F. (1997). *Information systems: An emerging discipline?* London: McGraw-Hill.

National Broadband Task Force. (2001). *The new national dream: Networking the nation for broadband access.* Canada: Ottawa Industry.

OECD. (2001). *Working party on telecommunication and information services policies: The development of broadband access in OECD countries.* Retrieved March 12, 2003, from OECD at http://www.oecd.org/dataoecd/48/33/2475737.pdf

Ofcom. (2004). *The Ofcom Internet and broadband update.* Retrieved November 20, 2005, from http://www.broadbanduk.org/reports/Ofcom percent20InternetandBroadband percent200104.pdf

Ofcom. (2005). *The Ofcom Internet and broadband update.* Retrieved November 20, 2005, from http://www.broadbanduk.org/reports/Ofcom percent20InternetandBroadband percent200404.pdf

Office of the e-Envoy. (2001). *UK online: The broadband future.* Retrieved July 15, 2003, from http://archive.cabinetoffice.gov.uk/e-envoy/reports-broadband/ $file/ukonline.pdf

Office of Technology Policy. (2002). *Understanding broadband demand: A review of critical issues.* U.S. Department of Commerce, Washington, DC.

Oftel. (2003). *International benchmarking study of Internet access (dial-up and broadband).* Retrieved November 30, 2003, from http://www.ofcom.org.uk/static/archive/Oftel/publications/research/2003/benchint_2_0603.htm

Oh, S., Ahn, J., & Kim, B. (2003). Adoption of broadband Internet in Korea: the role of experience in building attitude. *Journal of Information Technology, 18* (4), 267-280.

Rice, C. (1997). *Understanding customers.* Oxford: Butter worth-Heinemann.

Rogers, E. M. (1995). *Diffusion of innovations.* New York: Free Press.

Sawyer, S., Allen, J. P., & Heejin, L. (2003). Broadband and mobile opportunities: A socio-technical perspective. *Journal of Information Technology, 18* (4), 121-136.

Stanton, L. J. (2004). Factors influencing the adoption of residential broadband connections to Internet. In *Proceedings of the 37th Hawaii International Conference on System Sciences.*

Straub, D. W., Boudreau, M-C, & Gefen, D. (2004). Validation guidelines for IS positivist research. *Communications of the Association for Information Systems, 13,* 380-427.

Straub, D. W., Gefen, D., & Boudreau, M. C. (2005). Quantitative Research. In D. Avison & J. Pries-Heje, (Ed.), *Research in information systems: A handbook for research supervisors and their students.* Amsterdam: Elsevier.

Venkatesh, V., & Brown, S. (2001). A longitudinal investigation of personal computers in homes: Adoption determinants and emerging challenges. *MIS Quarterly, 25*(1), 71-102.

Chapter 2

Conceptual Model for Examining Consumer Broadband Adoption, Usage, and Impact

Abstract

An examination of previous literature in the information systems (IS) area illustrates that researchers have not yet undertaken research on broadband in the area of consumer diffusion, including the adoption, usage, and impact in the household (Crabtree, 2003; Oh et al., 2003; Stanton, 2004). Instead, most of the research associated with broadband has mainly focused on examining the macro level factors leading to adoption in a country (Crabtree, 2003; Oh et al., 2003; Stanton, 2004). Recently conducted studies highlight the need to understand adoption and diffusion of broadband from the household consumer perspective (Crabtree, 2003; Oh et al., 2003; Stanton, 2004). The limitation to studying adoption at a micro level has resulted in a lack of appropriate conceptual models specific to broadband. As pursued in previous adoption studies (Davis, 1989; Oh et al., 2003; Venkatesh & Brown, 2001), constructing a conceptual model specific to broadband diffusion at the household consumer level necessitates the review, identification, and integra-

tion of the relevant factors related with adoption, usage, and impact of technology previously examined in IS studies. Therefore, this chapter reviews and assesses the appropriateness of previous technology adoption models and constructs to study broadband diffusion. Then, this chapter provides further theoretical justification for selecting the constructs that are used to study broadband diffusion, formulate the hypotheses and finally draw a conceptual model of broadband diffusion. The chapter is structured as follows. The following section provides a review of the theoretical models of technology diffusion and adoption. This section also provides a brief discussion of the models applied to investigate broadband-related issues from the consumer perspective. Progressing upon this, the section thereafter briefly discusses the foundations of the proposed model and also provides an overall description of the proposed conceptual model. This is followed by an elaboration of the broadband diffusion model and the justification of the inclusion of the attitudinal, normative, control, behavioural intention, adoption behaviour, usage behaviour and impact constructs, and formulates the hypotheses by presenting theoretical explanations, past empirical findings, and practical examples. Finally, a summary of the chapter is provided.

Technology Diffusion and Adoption Theories

The study of adoption/acceptance, adaptations, and usage of information technology (IT) is considered to be one of the most mature areas of research within the IS discipline (Benbasat & Zmud, 1999; Hu et al., 1999; Venkatesh et al., 2003). Consequently, over time, a number of theories and models have been adopted from diverse disciplines such as social psychology, sociology, and marketing, and have been modified, developed, and validated by IS researchers in order to understand and predict technology adoption and usage (Benbasat & Zmud, 1999; Venkatesh et al., 2003). Theories and models that have been taken from other disciplines and developed by IS researchers include the theory of reasoned action (TRA) (Fishbein & Ajzen, 1975); the theory of planned behaviour (TPB) (Ajzen, 1991; 1988; 1985; Ajzen & Fishbein, 1980; Ajzen & Madden, 1986); the technology acceptance model (TAM) (Davis, 1989; Davis et al., 1989); and the diffusion of innovations (DI) theory (Rogers, 1995). According to the needs of IS research, these theories were further modified, extended and integrated. For instance, in order to understand various factors in detail, Taylor and Todd (1995) proposed the decomposed TPB by modifying TPB and integrating the diffusion of innovation constructs within it. Similarly, in order to understand the role of gender and social influence in technology adoption, Venkatesh and Morris (2000) extended TAM by integrating gender and subjective norm constructs with the original TAM model.

For the purpose of understanding technology adoption within the household context, Venkatesh and Brown (2001) modified the TPB to examine drivers and barriers of PC adoption. Due to the large numbers of choices of theories and models (e.g., TRA, TPB, DTP, TAM, DI, MATH), a selection of an appropriate model or various constructs from different models posed to be a problem for the upcoming technology adoption researchers. Venkatesh et al., (2003) argued that researchers are confronted with a choice amongst a multitude of models and find that they must "pick and choose" constructs across the models, or choose a "favoured model," and largely ignore the contributions from alternative models. This led Venkatesh et al. (2003) to review, discuss, and integrate elements across eight prominent user acceptance models (TRA, TAM, the motivational model, TPB, a model combining the technology acceptance model and the theory of planned behaviour, MATH, DI, and the social cognitive theory) that resulted in proposing the unified theory of acceptance and use of technology (UTAUT). UTAUT is composed of four core determinants (performance expectancy, effort expectancy, social influence, and facilitating conditions) of intention and usage. This model also consists of four variables such as gender, age, voluntariness, and experience that moderate key relationships between the aforementioned four core determinants and intention and usage.

Although the models and theories are widely tested and validated to explain the usage and adoption of technology from the 'users' perspective, their application is limited to studying the 'consumers.' Realising the potential of emerging ICTs for household consumers, recently, IS researchers have also begun to investigate the consumer adoption, usage and the impact of technology issues (Anckar, 2003; Oh et al., 2003; Venkatesh & Brown, 2001). Since the focus on consumers within the IS field is new (Venkatesh & Brown, 2001), the guiding theories, models, and research approaches are in the initial stage of development, testing, and validation. Therefore, in order to assist the selection of an appropriate model and/or constructs for current research, the following section will review the prominent technology diffusion and adoption models and will highlight their strengths and weaknesses to study the adoption and usage of technology from the consumers' perspectives. The theory and models will be discussed in light of the empirical studies available in the related area. A combination of the various models led this research to propose the model of consumer diffusion of broadband technology and this will be discussed in detail in subsequent sections.

Diffusion of Innovations

The diffusion of innovations theory describes the patterns of adoption, illustrates the process, and assists in understanding whether and how a new invention will be successful (Rogers, 1995). The meaning of the terms 'diffusion' and 'innovation' is already provided in a previous chapter (Chapter 1). This theory (Rogers, 1995; Tornatzky & Klein, 1982) has been employed to study a wide range of phenomenon including the adoption and usage of technology. Within the technology adoption and usage area, this theory was used to examine a variety of factors that are thought to be determinants of IT adoption and usage including individual user characteristics (Brancheau & Wetherbe, 1990), information sources and communication channels (Nilikanta & Scammell, 1990) and innovation characteristics (Hoffer & Alexander, 1992; Moore & Benbasat, 1991; Moore, 1987).

IS researchers also integrated the intentions and innovations theories combining concepts from the theory of reasoned action (Moore & Benbasat, 1991) and the theory of planned behaviour (Chau & Hu 2001; Taylor & Todd, 1995) with the perceived characteristics of innovations (Rogers, 1995). Since relative advantage as an innovation characteristic is employed in a number of studies (Moore & Benbasat, 1991; Tan & Teo, 2000; Taylor & Todd, 1995) and easily integrated with constructs from other theories and models, it is appropriate to consider this particular innovation attribute to examine the diffusion of broadband. The diffusion of innovations theory also suggests that demographic factors such as age, education and income represent correlates of innovativeness, which in turn determine the adoption rate of the innovation (Rogers, 1995). Therefore, this aspect of diffusion of innovations theory was considered relevant to investigate broadband adoption.

Theory of Planned Behaviour (TPB) and its Variations

Although the theory of planned behaviour (TPB) (Ajzen, 1985, 1988, 1991, Ajzen & Madden, 1986; Schifter & Ajzen, 1985) has its roots within social psychology, it is widely adopted and adapted by IS researchers to the study of IT adoption, implementation, and use (Benbasat & Zmud, 1999). It is an extended form of the theory of reasoned action (TRA), which was developed to overcome the TRA's limitations that dealt with an incomplete volitional control (Ajzen & Fishbein, 1980; Fishbein & Ajzen, 1975).

According to the TPB, human action such as an individual's adoption or use of a technology (i.e., broadband) is guided by the following three types of beliefs: first, behavioural beliefs that create a favourable or unfavourable attitude toward the behaviour; second, normative beliefs that produce perceived social pressure or subjective norms; and third, control beliefs that generate perceived behavioural control (Ajzen, 1988, 1991, 2002). These three types of constructs (i.e., attitude toward the behaviour, subjective norms, and perception of behavioural control) lead to the formation of a behavioural intention (Ajzen, 1988, 1991, 2002). The more favourable the attitude and subjective norm and the greater the perceived control, the stronger the person's intention should be to perform the behaviour in question (Ajzen, 1988, 1991, 2002). Finally, if consumers have strong actual control over the behaviour, they are more likely to execute their intentions in favourable circumstances. This suggests that intention is an immediate antecedent of behaviour

(Ajzen, 1988, 1991, 2002). In addition to intention, it is also useful to consider perceived behavioural control as a direct antecedent of behaviour. This is because to the extent that perceived behavioural control is stronger, it can serve as a substitute for actual control; hence, it can contribute to the prediction of the behaviour in question (Ajzen, 1988, 1991, 2002).

Although the TPB does not describe the process of implementation in a specific context, it has a high degree of predictive validity and can be used to identify areas of concern for a specific context (Benbasat & Zmud, 1999). According to the IS literature, the TPB can serve as an effective diagnostic tool when examining IT adoption or acceptance and usage (Benbasat & Zmud, 1999). Therefore, the TPB can be considered as a guiding framework when developing the proposed conceptual model for this research.

In order to increase the predictability of the TPB, Taylor and Todd (1995) decomposed the attitudinal belief dimensions and included the innovations characteristics (Rogers, 1995) as different dimensions of the attitude construct. Taylor and Todd's (1995) study concludes that a decomposed structure helps to increase predictability in comparison to the TPB and a more in-depth understanding when compared to TAM. This variant of the TPB is termed as the decomposed theory of planned behaviour (DTPB) (Taylor & Todd, 1995). Bearing this reasoning in mind, the conceptual model of broadband diffusion adopted the decomposed structure of attitude, subjective norms, and perceived behavioural control constructs. However, the constructs of these are not exactly similar to Taylor and Todd's (1995) study. This is because the context and subject of the two studies are different to one another.

Technology Acceptance Model (TAM)

The technology acceptance model is an adaptation of the theory of reasoned action. This model predicts systems usage by employing two factors, namely, the perceived usefulness and perceived ease of use of an IS (Davis, 1989). Perceived usefulness represents the beliefs of users, which is that technology use will enhance performance. These two factors determine the attitude towards the intention to use the system in question. Although TAM is a very successful model in terms of studying the users' intention of adoption and usage of technology, its application is yet to be investigated for consumers within the household context. Therefore, the TAM constructs were not considered, such as those proposed in the model of broadband diffusion.

Model of Adoption of Technology in Households (MATH)

The model of adoption of technology in the household (MATH) was applied to investigate PC adoption in American households (Venkatesh & Brown, 2001). According to MATH, technology adoption in the household is determined by a number of factors. These include the attitudinal belief structures such as utilitarian outcomes, hedonic outcomes, and social outcomes, normative belief structures such as the influence of friends, family, secondary information sources, and a control belief structure that consists of three barriers, namely knowledge, difficulty of use, and cost (Venkatesh & Brown, 2001).

The majority of the constructs included in this model are also useful to study broadband adoption. However, constructs from this model do not provide insights to the usage and impact of PCs: they only shed light upon the adoption of it (Venkatesh & Brown, 2001). Furthermore, this model was constructed to study PC adoption. As a result, detailed factors of MATH need to be adjusted for broadband. Theoretical justification for the selection of each construct is provided in the following sections.

Use Diffusion Model (UD)

The use diffusion model was developed to investigate technology use in the household context. The model was guided by the following three key components. First, use diffusion (UD) determinants such as household social context, technological dimension, personal dimension, and external dimension. Second, UD patterns which are the typology of uses or users consist of two constructs called the variety of use and rate of use. Third, UD outcomes consist of the perceived impact of technology, satisfaction with technology and interest in future technologies (Shih & Venkatesh, 2004). Since this model focuses upon the usage of technology in the household context, its constructs such as variety of use and rate of use would be useful to determine broadband usage.

Model Applied to Study Broadband Adoption and Diffusion

The discussion in Chapter 1 illustrated that the majority of studies conducted to understand broadband related issues are macro and exploratory in nature. These studies are mostly data driven in nature compared to process driven. Therefore, the use of conceptual models or theories is less prominent in initial studies (Anderson et al., 2002). However, as the broadband area has matured and adoption rates have increased, researchers have begun to use theory-based investigations to study the issue of diffusion (Oh et al., 2003), although such research is still in the embryonic stage.

Oh et al. (2003) integrated innovation attributes, such as compatibility, visibility, and demonstrability with technology acceptance constructs, such as perceived usefulness and perceived ease of use. Findings of this study suggest that these innovation attributes have a significant influence on the constructs in the extended technology acceptance model such as perceived usefulness, perceived ease of use, and perceived resources. On the basis of this study, the implications are that efforts should be made to expand the compatible experience base of broadband Internet in order to facilitate its adoption and use.

Although Oh et al.'s (2003) study is a good beginning, it is limited when examining broadband adoption behaviour, since it does not provide any evidence of how attitude building affects the behavioural intention to adopt or reject the technology in question. Also, the study examined the users of broadband rather than the consumers; therefore the findings are weak as they are only limited to the drivers and inhibitors of broadband adoption. This study identified the following two important limitations: first, the research model did not consider behaviour or behavioural intention constructs; second, the selection and use of statistical tools of data collection (Oh et al., 2003). Consequently, a study that will be a refinement of the research model and uses rigorous statistical tools and hypothesis testing was recommended (Oh et al., 2003).

A review of the technology adoption, diffusion theories and models suggest that MATH is the most appropriate and closest model to the context of this current study (i.e., technology adoption in the household consumer). MATH itself is based on the principle of the TPB and the DTPB, and is composed of constructs particularly suited to evaluate the usefulness of technology for household purposes. Therefore, MATH, TPB, and DTPB will be utilised as a guiding framework for developing a conceptual model that examines the behavioural intention to adopt broadband and actual broadband adoption and non-adoption. The diffusion of innovations theory suggests that socio-economic determinants, such as age, gender, income, education, and occupation, should be included within a study such as this. This allows a distinction to be drawn between broadband adopters and non-adopters. The review of the diffusion theories and models also suggest the appropriateness of the use of the diffusion model to study the usage and impact of broadband within the household context. Considering the limitations of the previously examined constructs for broadband adoption (Oh et al., 2003), the current study proposes to include the constructs, which are important to understand not only towards building attitudes towards adoption, but also the behavioural intention to adopt broadband, its relationship with the actual adoption, sustained adoption, usage, and impact. In other words, the conceptual model proposed in the following sections provides a holistic view of broadband diffusion within the household context.

Foundations of the Proposed Conceptual Model

Before describing the development of the proposed conceptual model for broadband adoption, usage, and impact, this section briefly discusses the underlying reasons for considering a guiding theory and model as a foundation for the proposed conceptual model.

Taylor and Todd (1995) identified two main criteria when selecting an appropriate model. First, a model that provides good predictions while using the fewest predictors is preferable; in other words, it is more parsimonious (Bagozzi, 1992; Taylor & Todd, 1995). Second, the model should provide reasonable predictive ability and should also contribute enough in providing an understanding of the phenomenon under investigation (Taylor & Todd, 1995). Since a broadband diffusion study requires both a predictive ability (in the case of adoption) and a contribution to understanding (in the case of usage and impact), the second criteria was adopted when developing the conceptual model. At a conceptual level, the constructs from the various models were included, which may in the future provide insights to an understanding of all the three stages of diffusion. However, after validation of these constructs, to maintain parsimony of the explanatory model the less significant constructs will be eliminated.

A number of related theories and models (such as TRA, TPB, DTPB, diffusion of innovations, and TAM) were described that are widely applied to study technology adoption from the individual or user's perspectives. However, it was also identified that their application is limited to study technology adoption from the household perspective (Venkatesh & Brown, 2001). As has been discussed, only the MATH model was successfully applied in the household context in order to examine PC adoption in the USA (Venkatesh & Brown, 2001). Therefore, as the household consumer is also the context of this research, MATH was utilised.

The previous discussion suggests the appropriateness of TPB, DTPB and MATH to study technology adoption issues from the perspective of the household context. Therefore, the conceptual model of broadband adoption (MBA) is an adaptation of MATH (Venkatesh & Brown, 2001) and is based on the DTPB (Taylor & Todd, 1995) and the TPB (Ajzen, 1985, 1991).

Consistent with the TPB, the proposed model of broadband adoption consists of three predictor types, namely attitudinal, normative, and control and dependent variables that include behavioural intention (BI) and adoption behaviour. Also, the relationship between the dependent and independent variables is hypothesised according to the TPB. This is because the TPB is a generalised theory and can be applied to a wide variety of contexts for predicting the adoption of different types of IT. Its major constructs reflect the key variables that have been identified as influential in previous implementation research and are flexible enough to subsume situation-specific factors (Benbasat & Zmud, 1999). Therefore, it is considered to be a basic guiding theory for this research. However, the TPB does not suggest breaking each of the predictor categories in more than one factor (Taylor & Todd, 1995); therefore it limits the researcher's ability to differentiate between the smaller factors within each category.

The DTPB, which is a variance of the TPB, allows researchers to decompose the attitudinal, normative, and control categories to better understand the reasons of adoption and non-adoption (Taylor & Todd, 1995). Therefore, the decomposed belief structure for household broadband adoption model is consistent with the DTPB.

Since MATH is directly focused upon investigating technology adoption from the household perspective, the majority of the detailed constructs for attitudinal, normative, and control categories are adapted from this (Venkatesh & Browns, 2001). At this point, this chapter will state why MATH in the original form is not employed to study broadband adoption. An adaptation of MATH was essential because the research objects of the two studies are dissimilar in nature. The research object of MATH was a PC, but for this chapter it is broadband. These two technologies are not similar in characteristics. Therefore MATH was subjected to the following modifications before it was applied for the examination of broadband adoption.

First, broadband has clear advantages over its predecessor narrowband. Therefore it was regarded as being important to examine the influences that are triggered by the advantages that broadband offers (such as faster access, always-on access, faster download, un-metered access, and free home phone line) over narrowband. Roger (1995) suggested the 'relative advantage' term for the perceived advantage of having a new technology over its predecessor. Taylor and Todd (1995) included the relative advantage attribute of Rogers' diffusion of innovations theory as an attitudinal construct to study the adoption of technology. Consistent with Taylor and Todd's (1995) model, this research also included relative advantage as an attitudinal construct.

Second, this research's model also included service quality as an attitudinal construct to evaluate a consumer's favourable or unfavourable perception of service quality. That is, a consumer could determine the quality of service being received from his/her current service provider and how that issue influences him/her to stay with the same provider in the long term or to switch to a new provider. This construct was not included in MATH. A PC (i.e., hardware) is a product that either does not require a seller to provide a consistent service after a product is sold or it does so to a minimal degree, and this could be a possible reason for not including 'service quality' in MATH.

Third, MATH considered only one construct 'social influence' within the normative category. But in the proposed model, social influence is differentiated by two types of influences, which are 'primary social influence' and 'secondary social influence.' This is consistent with Taylor and Todd's approach. The reason for the differentiation is due to the type of influence exerted by the two being different in nature; hence, the triggered influence is different and expected to have an impact on the various stages of adoption. Primary social influence is direct in nature and exerted by friends, peers, family members, and relatives who are expected to have a strong influence when performing certain behaviour. This is then followed by secondary social influence that is caused by the media. Therefore, it was considered to be appropriate to differentiate the social influence construct in the two abovementioned constructs within the normative category.

Fourth, data analysis for the MATH study only estimated the mean score and standard deviation. This was done in order to assess the importance of each factor; however, no test of statistical significance was reported. In contrast, the current model is based on several hypotheses that will be tested for their significance and will confirm the validity of the model and its predictive ability.

Finally, MATH did not examine if adoption behaviour differs for consumers in terms of the variables age, gender, income, education, and occupation. This research also has a scope to examine this issue.

All this juncture discussion so far provides reasoning for adapting constructs from MATH for developing the conceptual model for this study. The following sections describe the proposed model in detail and provide theoretical justification for the role of the various underlying constructs in determining household broadband adoption and usage behaviour.

Description of the Proposed Conceptual Model

The adoption component of the proposed diffusion model postulates that the behavioural intentions (BI) to adopt broadband are determined by the following three types of constructs (Ajzen, 1991; Rogers, 1995; Taylor & Todd, 1995; Venkatesh & Brown, 2001). These are: (1) **attitudinal constructs** (*relative advantage, utilitarian outcome,s and hedonic outcomes*) represent the consumers' favourable or unfavourable evaluation of the behaviour in question (i.e., adoption of broadband); (2) **normative constructs** (*primary and secondary influence*) represent the perceived social pressure to perform the behaviour in question (i.e., adoption of broadband); (3) **control constructs** (*knowledge, self-efficacy, and facilitating conditions resources*) represent the perceived control over the personal or external factors that may facilitate or constrain the behavioural performance (Ajzen, 1991; Rogers, 1995; Taylor & Todd, 1995; Venkatesh & Brown, 2001). The predictor variables from the three categories are expected to determine and explain the BI to adopt broadband, which in turn is expected to predict the actual broadband adoption behaviour (BAB) (Ajzen, 1991; Rogers, 1995; Taylor & Todd, 1995; Venkatesh & Brown, 2001).

Additionally, this research also proposes that the BAB can also be explained by the demographic characteristics of the adopters and non-adopters in terms of age, gender, education, income, and occupation. The usage and impact components of the proposed conceptual model puts forward that the BAB determines the: (1) rate of Internet use (i.e., the total time spent online and the frequency of Internet access), and (2) variety of use (i.e., activities performed online). The impact of broadband is postulated as an ultimate outcome of usage of broadband (Shih & Venkatesh, 2004; Vitalari et al., 1985). The TPB, the DTPB and MATH are utilised as guiding theories

Table 2.1. Summary of research hypotheses

HN	Independent Variables	Dependent Variables
H1	Overall Attitudinal Constructs	Behavioural Intention to adopt broadband (BI)
H2	Relative Advantage	Behavioural Intention to adopt broadband (BI)
H3	Utilitarian Outcomes	Behavioural Intention to adopt broadband (BI)
H4	Hedonic Outcomes	Behavioural Intention to adopt broadband (BI)
H5	Service Quality	Behavioural Intention to change service providers (BISP)
H6	Overall Normative Constructs	Behavioural Intention to adopt broadband (BI)
H7	Primary Influence	Behavioural Intention to adopt broadband (BI)
H8	Secondary Influence	Behavioural Intention to change service providers (BISP)
H9	Overall Control Constructs	Behavioural Intention to adopt broadband (BI)
H10a	Facilitating Conditions Resources	Behavioural Intention to adopt broadband (BI)
H10b	Facilitating Conditions Resources	Broadband Adoption Behaviour (BAB)
H11	Knowledge	Behavioural Intention to adopt broadband (BI)
H12	Self-efficacy	Behavioural Intention to adopt broadband (BI)
H13	BI	Broadband Adoption Behaviour (BAB)
H14	Age	Broadband Adoption Behaviour (BAB)
H15	Gender	Broadband Adoption Behaviour (BAB)
H16	Education	Broadband Adoption Behaviour (BAB)
H17	Income	Broadband Adoption Behaviour (BAB)
H18	Occupation	Broadband Adoption Behaviour (BAB)
H19a	Duration	Broadband Adoption Behaviour (BAB)
H19b	Frequency	Broadband Adoption Behaviour (BAB)
H19c	Variety of Use	Broadband Adoption Behaviour (BAB)

in order to determine the relationships amongst the various constructs in the proposed conceptual model that will investigate the following research questions:

RQ 1: Do attitudinal (relative advantage, utilitarian outcomes, and hedonic outcomes), normative (primary influence) and control factors (knowledge, self-efficacy and facilitating conditions resources) influence behavioural intentions when adopting broadband?

RQ 2: How strongly do attitudinal (relative advantage, utilitarian outcomes, and hedonic outcomes), normative (primary influence) and control factors (knowledge, self-efficacy, and facilitating conditions resources) influence behavioural intentions when adopting broadband?

RQ 3: Do behavioural intentions and control factors influence the actual adoption of broadband?

RQ4: Do demographic factors (i.e., age, gender, education, income, and occupation) influence the adoption of broadband?

RQ 5: Do service quality and secondary influence affect the behavioural intentions when changing current service provider?

RQ 6: Does the rate of Internet usage differ for the broadband and narrowband users?

RQ 7: Do broadband users access more online activities than narrowband users?

RQ 8: Does the use of broadband affect the time spent on various daily life activities?

A list of constructs, summaries of hypotheses are presented in Table 2.1. A definition of each construct is provided in Table 2.2. The following sections also provide detailed descriptions of each construct and the theoretical justification for including them in the proposed conceptual model.

Attitudinal Constructs

Attitude is defined as an individual's positive or negative feelings when performing target behaviour, such as adoption of broadband (Ajzen, 1985, 1991; Fishbein & Ajzen, 1975; Taylor & Todd, 1995). Overall, the technology adoption/acceptance theories and models including the TRA, TAM, TPB, and DTPB illustrate relationships between attitude or attitudinal factors and behavioural intentions. If the attitude of individuals towards the technology in question is positive, then they are likely to form an intention to perform the behaviour (Tan & Teo, 2000). Following the TPB, it can be assumed that if the perception of the respondents regarding the attitudinal factor is positive, then it is more likely that it will have a positive influence on the behavioural intention. Thus, this leads to the formulation of the hypothesis:

Table 2.2. Definition of constructs

Constructs	Definitions of constructs and sources
Behavioural Intention (BI)	Behavioural Intention (BI) is defined as a consumer's intention to subscribe (or intention to continue the current subscription) and makes use of Broadband Internet in the future. (Ajzen, 1988; 1991; Taylor & Todd, 1995; Venkatesh & Brown, 2001).
Relative Advantage	It is defined as the degree to which broadband Internet is perceived as being better than its predecessor narrowband Internet. (Moore & Benbasat, 1991; Rogers, 1995; Tornatzky & Klein, 1982).
Utilitarian Outcomes	It is the extent to which broadband Internet usage enhances the effectiveness of household activities such as, undertaking office work at home, children's homework, information or product search and purchase and home business (Brown & Venkatesh, 2003; Venkatesh & Brown, 2001).
Hedonic Outcomes	Hedonic outcomes are defined as the pleasure derived from the consumption, or use of broadband Internet. For example, the entertainment potential of the Internet via offerings such as, online radio, streaming audio and video, electronic greetings, online games, online casino (Brown & Venkatesh, 2003; Venkatesh & Brown, 2001).
Service Quality	Service quality can be defined as the perceived quality of service a consumer obtained or is obtaining from the current Internet service providers. Service quality is measured in terms of, speed of connection and security problem with Internet connections, virus and popup problems with connection and customer support obtained from the ISP providers (DeLone and McLean, 2003; Parasuraman *et al*, 1991; Parasuraman *et al*, 1991).
Primary influences	Primary influences are defined as the perceived influences from friends and family to subscribe to and use (or not to subscribe and use) broadband Internet services (Brown & Venkatesh, 2003; Venkatesh & Brown, 2001).
Secondary Influences	Secondary influences are defined as the perceived influence of information from secondary sources such as advert and news on TV, newspapers to subscribe and use (or not to subscribe and use) broadband Internet services (Rogers, 1995; Brown & Venkatesh, 2003; Venkatesh & Brown, 2001).
Knowledge	Knowledge is defined as the perceived level of knowledge about broadband Internet, its risks and benefits (Rogers, 1995; Venkatesh & Brown, 2001).
Self-efficacy	Self-efficacy is defined as the perceived ability or skill to operate computers and the Internet (narrowband or broadband) without the assistance of others (Ajzen, 1985, 1991; Taylor & Todd, 1995).
Facilitating Conditions Resources	Facilitating conditions resources is defined as the perceived level of resources when subscribing to broadband (Ajzen 1985; 1991; Taylor & Todd, 1995).

H1: *Overall attitudinal factors will have a positive influence on the behavioural intention to adopt broadband.*

In order to gain in-depth and better understanding, the attitude construct has been decomposed in several studies (Tan & Teo, 2000; Taylor & Todd, 1995; Venkatesh & Brown, 2001). For example, Taylor and Todd (1995) decomposed attitude into five constructs represented by the five perceived innovation attributes (i.e., relative advantage, compatibility, visibility, risk, and complexity) from Rogers' diffusion of innovations theory. Following Taylor and Todd, Venkatesh and Brown (2001) also decomposed the attitudinal belief to study the adoption of the PC into three types of constructs, namely, utilitarian outcomes, hedonic outcomes, and social outcomes.

Following the previous discussion (Taylor & Todd, 1995; Venkatesh & Brown, 2001) this research decomposed attitude into four constructs: hedonic outcomes, utilitarian outcomes (Venkatesh & Brown, 2001), relative advantage (Rogers, 1995) and service quality (DeLone & McLean, 2003; Parasuraman et al., 1991).

Three constructs—relative advantage, utilitarian outcomes, and hedonic outcomes—are expected to provide measures of attitude towards the behaviour of broadband adoption in the household. Apart from these constructs, this study also included service quality as one construct of attitude for those consumers who already had broadband. This construct will be helpful to predict if the adopters are contracted or obligated to the same broadband provider. Alternatively, it will help predict whether the adopters switch to another provider if they are not satisfied with the obtained service. The attitudinal factors relevant to this study are discussed below in detail. Following this, the related hypotheses are formulated.

Relative Advantage

Rogers' (1995) diffusion of innovations theory suggests that the perceived relative advantage of an innovation is positively related to its rate of adoption. Several previous empirical studies have found that perceived relative advantage is an important factor for determining the adoption of an innovation (Tan & Teo, 2000; Taylor & Todd, 1995; Tornatzky & Klein, 1982). Similarly, when compared to narrowband, broadband offers faster, un-metered, always-on access to the Internet, and provides a number of advantages, conveniences and satisfaction to its users. Considering the advantages that broadband offers, it would be expected that individuals who perceive broadband as advantageous would also be more likely to adopt the technology. Therefore, the above theoretical argument leads to the following hypothesis:

H2: *Relative advantage will have a positive influence on behavioural intention.*

Utilitarian Outcomes

The perceived usefulness construct (Davis, 1989) is one of the strongest predictors used to examine the adoption and usage of workplace technology. Venkatesh and Brown (2001) proposed and validated the utilitarian outcomes factor that can be used to examine the adoption and usage of technology in a household setting. Utilitarian outcomes are the extent to which using a PC enhances the effectiveness of routine, household activities, such as budgeting, homework, and work (Venkatesh & Brown, 2001). It has been suggested that broadband can offer a more flexible lifestyle (BSG, 2004). For instance, many people subscribe to broadband in order to work at home instead of travelling to the office; broadband can assist children with their homework, and many more household activities can be performed conveniently using the faster access of the Internet offered via broadband. Therefore, it is expected that the greater the perception of broadband's usefulness for work or household related activities, the more likely that broadband technology will be adopted in the home. Thus, the following hypothesis is generated:

H3: *Utilitarian outcomes will have a positive influence on behavioural intention.*

Hedonic Outcomes

The Venkatesh and Brown (2001) study found that hedonic outcomes is one of the factors that influences PC adoption in the home. Venkatesh and Brown (2001) define hedonic outcomes as pleasure derived from PC use; for example, games, fun, and entertainment. Heijden (2004) described hedonic information systems as self-fulfilling and strongly connected to the home and leisure activities, focused on the fun aspect of using information systems, and encouraged prolonged rather than productive use (Heijden, 2004). Empirical findings from Venkatesh and Brown's (2001) study established that when adopting a technology, the role of entertainment (PC games, video games) was an important factor of consideration in the consumer decision-making process (Venkatesh & Brown, 2001). Previous studies suggest that the entertainment potential of a PC offers a possibility to escape reality and become immersed in a new environment. Such characteristics are consistent with a hedonic perspective (Foxall, 1992; Venkatesh & Brown, 2001). Hence, the entertainment potential of a PC is much more enhanced by the advent of the Internet. It offers the opportunity of playing online games, downloading music and video, chat and sending online messages. However, this potential was severely hampered by the slow speed of dial up Internet.

The barrier of slow speed was overcome by broadband technology, which offered benefits in terms of data, faster download speeds, and streaming capabilities to Internet users, resulting in more convenient and compelling environments. Recent studies (Lee et al., 2003, Lee & Choudrie, 2002) suggest that one of the most important factors responsible for broadband adoption in South Korea was the PC bang phenomenon. Similarly, a study by Anderson et al., (2002) suggests that broadband users are more likely to use the Internet for fun and entertainment in comparison to narrowband users. Considering the entertainment potential that broadband offers, it is expected that individuals who perceive broadband as a good entertainment medium will also be likely to adopt the technology. Therefore, the underlying hypothesis is:

H4: *Hedonic outcomes will have a positive influence on behavioural intention.*

Service Quality

Marketing research developed the construct service quality 'SERVQUAL' in order to measure a consumer's perception of service quality (Parasuraman et al., 1988; Parasuraman et al., 1991). However, only a limited number of studies have recently included it to measure the successful adoption of technology. DeLone and McLean (2003) extended a decade-old IS success model (DeLone & McLean, 1992) by integrating a new construct called 'service quality.' Service quality was included to evaluate the fact that an IS department also plays a role in facilitating end-user computing via the services it offers to business personnel wishing to develop their own systems (Rosemann & Vessey, 2005). McCalla and Ezingeard (2005) have progressed to develop a data collection protocol that measures the relationship between technology use, emotional expression and service quality perception. Yang et al., (2005) developed and validated an instrument that measured a user's perceived service quality of information presented on web portals. Parasuraman et al., (2005) have recently developed a multiple-item scale for assessing electronic service quality. Therefore, it can be deduced from these studies that there is a growing need and importance of the service quality construct within IS research.

However, despite its apparent significance, the service quality construct was not employed in the case of PC adoption studies. This is because when purchasing PCs (i.e., hardware) consumers have only one opportunity to make a choice; that is, to purchase or not to purchase. And once a PC is sold to a consumer, the seller is least expected to provide any further after purchase service and/or customer support. However, the case of broadband subscription is different to PC purchase. That is, the consumers sign an annual contract and during this period, if the provided service is not satisfactory, they can/will discontinue the broadband subscription. Alternatively, if consumers have a choice of providers they might transfer to the competitors. Therefore, it is important to understand whether consumers are satisfied with their current providers and provided services. Hence, the underlying hypothesis is:

H5: *Service quality will have a negative influence on the behavioural intention when changing from a current service provider.*

Normative Constructs

Subjective norms are defined as a consumer's perception that most people who are important to him/her think that s/he 'should or should not perform the behaviour in question' (Ajzen, 1985, 1991; Fishbein & Ajzen, 1975, Tan & Teo, 2000; Taylor & Todd, 1995; Venkatesh & Brown, 2001). A subjective norm in its original form in the TRA and TPB is employed as a single construct and is considered directly related to the behavioural intention. This is because a consumer's behaviour is based on their perception of what others think of what they should be doing (Tan & Teo, 2000). Following the guidelines of the TPB, it can be assumed that the stronger the perceived social influence, the more likely that the consumer develops a stronger intent to subscribe to broadband. Thus, this leads to the hypothesis:

H6: *Overall, the normative factors will have a positive influence on the behavioural intention when adopting broadband.*

Following the suggestion from various studies (Burnkrant & Page, 1988; Oliver & Bearden, 1985; Shimp & Kavas, 1984), Taylor and Todd (1995) decomposed the normative belief structure into two groups: the peer and superior influences. In terms of consumer-oriented service, relevant references, such as the adopter's friends, family, and colleagues/peers may influence the adoption decision (Tan & Teo, 2000). Rice et al. (1990) defined such influence as social influences where members of a social network influence others' behaviour. Venkatesh and Brown (2001) have considered the social influence of family, friends, TV, and newspapers as one construct that can be used to measure the subjective norms. The findings of Venkatesh and Brown (2001) suggest that social influences are significant determinants of the purchasing behaviour of PCs. Similarly, it is expected that households with broadband connections are likely to influence their relatives and friends by telling and demonstrating to them the benefits and convenience offered by broadband. Therefore, it is appropriate to consider social influences as a measure of the subjective norm for broadband adoption in the household. By using previous studies (Fulk & Boyd, 1991; Fulk et al., 1987; Salancik & Pfeffer, 1978), Venkatesh and Brown (2001) suggested that 'social influence is exerted through messages and signals that help to form perceptions of a product or activity.' Measures that influence adopters can appear in two forms that are termed as primary and secondary influences. Consistent with the DTPB and MATH, this study considered two constructs in the normative category. These two constructs are separated and defined below. Subsequently, the underlying hypotheses are proposed.

Primary Influences

For the purposes of this research, social influence from friends, colleagues/peers, and family members that takes the form of a conversation, messages, and assists in forming perceptions of broadband adoption is considered to be a primary influence (Venkatesh & Brown, 2001). Considering the findings from previous studies (Venkatesh & Brown, 2001; Taylor & Todd, 1995), this research assumes that if broadband adopters are influenced by their social networks with positive messages, they are more likely to have a strong behavioural intention to adopt broadband. Thus, the hypothesis proposed from this discussion is:

H7: *Primary influences will have a positive influence on the perceived behavioural intention to adopt broadband.*

Secondary Influences

Previous studies suggest that messages disseminated using mass media, such as the television (TV) and newspaper advertisements (secondary sources of information) are likely to influence an adopter's intentions (Rogers, 1995; Venkatesh & Brown 2001). For the purposes of this research, it is expected that secondary sources of information will affect those consumers who have already adopted broadband but are not satisfied with service quality. Thus, if advertisements are viewed on TV or read in a newspaper about broadband packages that are economical and offer a better quality service, then they are more likely to cause adopters to contract with the new provider. Therefore, the hypothesis is:

H8: *Secondary influences will have positive influence on perceived behavioural intention to change current service providers.*

Control Constructs

The TPB suggests that presences of constraints can inhibit both the behavioural intention to perform behaviour and the behaviour in question itself (Ajzen, 1991, 1985), which is referred to as the perceived behavioural control (PBC). Support for this theoretical argument is obtained from empirical findings in several studies that illustrate that the higher the perception of an individual's control over their internal and external constraint, the more likely that s/he will adopt the technology in question (Ajzen, 1991; Tan & Teo, 2000). However, if the individual's control over the external and internal constraints is low, then besides having a strong behavioural intention, s/he is less likely to adopt the technology (Ajzen, 1991, 1985). Consistent with the TPB, this research therefore formulates the following hypothesis:

H9: *The overall control factors will have a positive influence on the behavioural intention to adopt broadband.*

To develop a better understanding, subsequent studies have decomposed PBC into three constructs: knowledge, facilitating conditions resources/technology (Taylor & Todd 1995; Mathieson, 1991), and self-efficacy (Taylor & Todd 1995). To study PC adoption in the household, Venkatesh and Brown (2001) split PBC into five specific barriers that can inhibit the adoption of PC, including a rapid change in technology, declining costs, the high cost of PCs, ease/difficulty of use, and a requisite knowledge of the use of PCs.

Since the subscription cost of broadband access is stable and technology is not changing rapidly, the declining cost and rapid changes in technology were considered irrelevant factors for the adoption of broadband technologies, and are therefore not included in this research. Consistent with the DTPB and MATH, the current study considered the constructs that are barriers to the adoption of broadband as measuring high costs (i.e., facilitating conditions resources), the ease/difficulty of PCs and Internet use (i.e., self-efficacy) and the lack of knowledge of broadband's benefits. In the next section, the justification for including these three constructs as a decomposed control constructs and the hypotheses are provided.

Facilitating Conditions Resources

The South Korean government's vision recognised an affordable monthly cost of broadband for middle-income households as one of the most important factors that led to the high rates of adoption (Choudrie & Lee, 2004; Lee & Choudrie, 2002). An exploratory study on broadband adoption in the UK also suggests that a high monthly cost is a major barrier that is inhibiting the adoption of broadband in the household (Dwivedi et al., 2003). Therefore, it is expected that if the perceived cost of obtaining broadband is high, then adoption will be slow. Broadband technology is not compatible to the specifications of old PCs and necessitates either an upgrade or purchase of a new PC. However, PCs are not easily replaceable devices for the medium and lower income households. Therefore, an economic barrier in the form of costs that are incurred when upgrading or purchasing new personal computers inhibits the adoption of broadband in the household. Therefore, the hypotheses are:

H10a: *Facilitating conditions resources will have a positive influence on the behavioural intention to adopt broadband.*

H10b: *Facilitating conditions resources will have a positive influence on the adoption of broadband.*

Knowledge

The level of knowledge regarding an innovation, its risks, and benefits affect the adoption rate (Rogers, 1995). The greater the awareness of the benefits of the innovation amongst the consumers and users, the more likely it is that the innovation will get adopted. Lee and Choudrie's (2002) research suggested that in South Korea consumers knew what the potential of broadband was. The consumers were aware of the benefits of faster Internet access, which was essential to satisfy their needs. It is assumed that the adoption of broadband requires a clear message of its usage and benefits amongst the overall population. Also, if consumers are not aware of the benefits of adopting a particular innovation, then it is expected that they are more likely to reject the decision to make a purchase due to a lack of the perceived needs. Therefore, the underlying hypothesis is:

H11: *Knowledge will have a positive influence on the behavioural intention to adopt broadband.*

Self-Efficacy

Since the use of broadband also requires using a PC and the Internet, the ease or difficulty of use and requisite knowledge of a PC and Internet use are expected to have an impact upon broadband adoption. The South Korean government deployed a variety of promotion policies (Choudrie & Lee, 2004; Lee & Choudrie, 2002). "The Ten Million Program" was designed to boost Internet usage amongst housewives, the elderly, military personnel, farmers ,as well as excluded social sectors such as low-income families, the disabled, and even prisoners. This promotion of providing PC and Internet skills in the year 2000 contributed towards the adoption of the Internet. A total of 4.1 million new online users, including one million housewives, occurred as a result of such initiatives (Choudrie & Lee, 2004; Lee & Choudrie, 2002). Therefore, it is expected that household users with basic PC and Internet skills are more likely to adopt broadband. Hence, the hypothesis is:

H12: *Self-efficacy will have a positive influence on the behavioural intention to adopt broadband.*

Demographic Variables

Key socio-economic variables such as age, gender, education, income, and occupation (Burgess, 1986) provide important information regarding the characteristics of the population under investigation. These socio-economic variables have been widely applied to investigate a number of devices and issues within the IS discipline including the computer, telephone, Internet, software and e-learning technologies. For example, these variables have been included in previous studies that examined the adoption of ICTs such as the computer (Al-Jabri, 1996; Carveth & Kretchmer, 2002; Venkatesh et al., 2000; Vitalari et al., 1985), the telephone (Anderson et al., 1999), the Internet (Anderson & Tracey, 2001; Carveth & Kretchmer, 2002), and broadband (Anderson et al., 2002) in households, and its subsequent impact on users. Further, such social variables have also been applied to investigate software piracy (Solomon & Brien, 1990; Wood & Glass, 1995), technology adoption (Chen et al., 2001; Harris et al., 1996; Morris & Venkatesh, 2000; Venkatesh et al., 2003; Venkatesh & Morris, 2000), e-government adoption (Huang et al., 2002), and demographic differences amongst IS professionals (Holmes, 1997). Additionally, the previous studies also emphasised the role of several external variables, such as demographic characteristics on the decomposed belief structure and, ultimately, adoption, and usage (Venkatesh & Brown, 2001). The moderating effects of the demographic variables have been successfully applied to previous studies that examined gender differences in the perception and use of e-mail (Gefen & Straub, 1997), and relationships between organisational culture and computer efficacy (Pearson et al., 2002).

Since home computers, the telephone, and access to the Internet (both dial up and broadband) can be placed in the same technology cluster (Rogers, 1995), the socio-economic variables that have been employed to study one technology can also be used to study others (Rogers, 1995). Therefore, the socio-economic variables such as age, gender, education, occupation, and income that were utilised to examine home computer adoption in households can also be employed to study broadband adoption.

The socio-economic variables (i.e., age, gender, income, education, occupation) have also been widely studied within the marketing discipline. There are other demographic and geographic variables such as disability, ethnicity, marital status, and geographic locations that may provide useful information (Rice, 1997) on adoption. However, due to feasibility reasons (face-to-face interviews would have been required, but this is beyond the scope of this research) these variables were not included in this study. The study of the variables was termed as segmentation, which involves the breakdown of the total broad and varied markets into homogenous, distinct, accessible, stable and large groups (Gilligan & Wilson, 2003; Rice, 1997). Segmentation serves the following two important functions in marketing: first, to target marketing messages to appropriate segments, and second, to develop modified products that fit specific segments of the market (Gilligan & Wilson, 2003; Rice, 1997). Therefore, a study of the demographics of broadband consumers may assist the policy makers and ISPs by identifying the various segments' specific needs and constraints.

Age

Finch (1986) argued that age can be employed as a factor or independent variable to explain a particular social grouping, social process, or piece of individual or collective behaviour. Within the IS area, a number of studies have found evidence that explains the significant, direct, and moderating effect of age on the behavioural intention, adoption, and usage behaviours (Harris et al., 1996; Morris & Venkatesh, 2000; Pearson et al., 2002; Venkatesh et al., 2003). A study by Venkatesh et al., (2000) suggests that the majority age group adopting computers in the USA is 15 to17 years, which is then followed by the 26 to 35 years. Similarly, Lee and Choudrie (Lee et al., 2003; Lee & Choudrie, 2002; Choudrie & Lee, 2004) found that in South Korea, the group that increased the adoption of broadband via the PC bangs, was the younger age category (i.e., school attending age). In turn, the younger generation's usage of broadband exerted a substantial influence on the parents' decisions for subscribing to broadband as parents considered broadband to be imperative for educational and entertainment purposes. Carveth and Kretchmer (2002) found that in many West European countries, the older demographic groups are less likely to use the Internet compared to the younger ones. According to this study, in the UK, 85 percent of those aged 16-24 have Internet access compared to just 15 percent in the 65-74 age range, and 6 percent over the age of 75 (Carveth & Kretchmer, 2002). A study by Anderson et al. (2002) also suggests that the demography of dial up users is different to the broadband one. Therefore, significant age differences are expected in terms of broadband adopters and non-adopters. The younger and middle age groups are expected to be more apathetic to adoption, whilst the older age group is expected to be more relevant to the non-adopters. Hence, this forms the following hypothesis:

H14: *There will be a difference between the adopters and non-adopters of the various age groups.*

Gender

According to Morgan (1986) gender, as a key variable, is one of the most common variables in social investigations. Jackson and Scott (2001) defined gender as a hierarchical division between women and men embedded in both social institution and social practices. Gender is therefore a social structural phenomenon, but is also produced, negotiated, and sustained at the level of everyday interaction (Jackson &

Scott, 2001). Morgan (1986) argued that gender can be employed as a descriptive variable as well as an explanatory variable (Morgan, 1986). A number of studies have investigated the role of gender in the adoption and usage of ICTs (Al-Jabri, 1996; Anderson et al., 1999; Choudrie & Lee, 2004; Gefen & Straub, 1997; Harris et al., 1996; Holmes, 1997; Lacohee & Anderson, 2001; Morris & Venkatesh, 2000; Pearson et al., 2002; Venkatesh et al., 2003; Venkatesh & Morris, 2000; Venkatesh et al., 2000; Wood & Glass, 1995). The findings of the previous studies revealed that gender has an important role when considering technology adoption and usage in both the organisational and household contexts. A study by Venkatesh et al., (2000) illustrated that male users use a computer more than females and suggested gender as one of the most important variables when examining PC adoption in the household. Anderson et al., (1999) also suggest that clear gender differences exist in the usage of computers and telephone calls. Lacohee and Anderson (2001) also emphasised the differences between men and women in terms of the usage of a telephone. Hence, the hypothesis is:

H15: *The adopters of broadband will be more from male than female gender.*

Education

Previous studies suggest that individuals with educational qualifications are more likely to adopt new innovations (Burgess, 1986; Rogers, 1995). Past research on technology (PC) adoption suggests a positive correlation between the level of education, technology ownership, and usage (Venkatesh et al., 2000; Vitalari et al., 1985). Venkatesh et al., (2000) found that people with higher educational qualifications use computers more than the less educated ones. Education has also been identified as an important driver of broadband adoption in South Korea (Choudrie & Lee, 2004; Lee et al., 2003). Anderson et al. (2002) suggest that household consumers with secondary or tertiary education are more likely to have Internet access. From the previous research undertaken both in theory and empirical research, it is suggested that education can be considered as an independent variable that explains the differences between broadband adopters and non-adopters. This is because broadband is considered to be useful for educational purposes and performing job-related tasks. Therefore, it is expected that household consumers with higher educational attainment or those working towards higher educational attainment, i.e., degrees or postgraduate degrees, are more likely to adopt broadband. Hence, the derived hypothesis is:

H16: *There will be a difference between the adopters and non-adopters of broadband in different levels of education.*

Income and Occupation

As in the case of education, Rogers (1995) described socio-economic status (income and occupation) as a correlate or antecedent of innovativeness. The diffusion of innovations theory suggests that new technologies are initially adopted by those with more resources (Rogers, 1995). The adaptive structuration theory found that IT has the potential to increase the resources of both those who had them prior to its adoption and those who possessed fewer resources prior to its adoption (DeSanctis & Poole, 1994; Mason & Hacker, 1998). The findings of a longitudinal study using the USA census data found a positive correlation between income and computer ownership (Venkatesh et al., 2000). This study further suggests that a considerable gap persists between the lower and higher income groups (Venkatesh et al., 2000). A study by Anderson et al. (1999) also confirmed that income and occupation drive the general pattern of ICT ownership and usage. Anderson et al. (1999) found a strong correlation between social class and ownership of PCs, telephones and television. Similarly, Carveth and Kretchmer (2002) suggested that in the USA, the higher income families and households are more likely to own a computer and use the Internet. A similar pattern was suggested for Western European countries including the UK. Carveth and Kretchmer's study suggested that only 23 percent of lower income groups in comparison to 68 percent of the higher income groups in the UK use the Internet (Carveth & Kretchmer, 2002). A recent study focusing upon the determinants of the global digital divide also confirmed the importance of per capita income in explaining the gap between computer and Internet use (Chinn & Fairlie, 2004). The theoretical arguments and empirical evidence support the inclusion of both income and occupation as an independent variable that explains the difference between broadband adopters and non-adopters in households. Thus, the hypothesis is:

H17: *There will be a difference between the adopters and non-adopters of different levels of household annual income.*

Similar to income, occupation is also likely to differentiate between the adopters and non-adopters of broadband. This is because broadband is useful for performing job-related tasks; therefore, respondents with higher skilled occupation categories such as 'A,' 'B,' and 'C1' are more likely to adopt broadband, which is not expected for the lower occupation categories such as 'C2' and 'D,' but not including category 'E.' These occupation categories were derived from the marketing literature where mainstream professionals such as doctors, lawyers, and judges with the responsibility of more then 25 staff are classified as occupational category 'A' (Gilligan & Wilson, 2003; Rice, 1997). The occupations with a responsibility of less then 25 staff and academics are grouped as social grade 'B.' Skilled-non-manual workers fall within the occupational category 'C1' and 'C2.' Unskilled manual workers belong to the occupational category

'D.' Finally, housewives, retired individuals, students and unemployed citizens were placed in category 'E' (Gilligan & Wilson, 2003; Rice, 1997). As broadband provides a function to students and unemployed people who are engaged in job hunting, these groups are more likely to adopt broadband although they belong to the lower occupation category 'E.' This led to the formulation of the following hypothesis:

H18: *There will be a difference between the adopters and non-adopters of different types of occupation.*

Dependent Variables: Behavioural Intentions (BI) and Broadband Adoption Behaviour (BAB)

The TPB (Ajzen, 1991) considered two independent variables, namely BI and behaviour in question. The majority of technology adoption and usage research has utilised these two dependent variables to predict technology adoption and usage (Ajzen, 1991; Davis, 1989; Venkatesh & Brown, 2001). The TPB and findings from previous empirical studies suggest BI is a mediating variable between the predictors and actual behaviour. Therefore, BI is considered to have a direct influence on adoption or usage (Ajzen 1991). Apart from BI, previous studies have also employed a control factor (i.e., available resources) as a direct predictor of behaviour (Ajzen, 1991; Venkatesh & Brown, 2001). Previous studies have reported a strong correlation between control factor and behaviour (Ajzen, 1991). Findings from a number of technology adoption and usage studies within the IS field suggest BI (Morris & Venkatesh, 2000; Venkatesh et al., 2003; Venkatesh & Brown, 2001; Venkatesh & Morris, 2000; Venkatesh et al., 2000) and control (Taylor & Todd, 1995) as being good predictors of actual adoption or usage behaviour. Consistent with previous studies and the guiding theory, the current study considered BI as a mediating dependent variable and adoption behaviour as an ultimate dependent variable. The following hypothesis illustrates the relationship between BI and BAB.

H13: *Behavioural intention and facilitating conditions resources will have an influence on the adoption of broadband.*

Usage of Broadband

The proposed conceptual model of this research considered a variety of broadband use and rate of use as dependent variables. It is expected that the independent variable broadband adoption behaviour will differentiate between the variety and rate (Shih & Venkatesh, 2004) of Internet use between broadband and narrowband users. Following the previous study on technology usage (Shih & Venkatesh, 2004), we postulate the following hypotheses on broadband use:

H19a: *The adopters of broadband will spend more time online than non-adopters.*

H19b: *The adopters of broadband will access the Internet more frequently than non-adopters.*

H19c: *The adopters of broadband will access a higher number of online activities than the non-adopters.*

Impact of Broadband

Time allocation patterns are considered to be an important variable for understanding the role of computing and the impact of ICTs in households (Vitalari et al., 1985). This is due to the total available time being finite and the time spent using a technology being likely to influence the distribution of time for other activities. Therefore, the use of a new technology does indirectly rearrange the social actions and user behaviour (Vitalari et al., 1985). Vitalari et al. (1985) have demonstrated the impact of personal computers upon the time allocation of various routine activities and have also considered the implications of this action. Although broadband is expected to affect several aspects of daily life in the household (BSG, 2004; Carriere et al., 2000), research that understands the impact of broadband on time allocation patterns has not been undertaken. This was a motivating factor for including this construct.

According to the diffusion literature, new innovations such as broadband are likely to change the associated behaviours of users, which are termed as the perceived consequences or impact of new innovations (Rogers, 1995; Shih & Venkatesh, 2004). Previously, researchers have demonstrated the impact of various technologies (e.g., automobiles, telephone, computers, and Internet) on a user's daily life (Anderson et al., 2001; Vitalari et al., 1985). Since broadband offers an alternative way of work and entertainment, and consumes time that traditionally has been spent on other activities, it is likely that broadband will alter the time allocation pattern of a user's daily activities.

The homeostasis relationship between technology and time change (Robinson, 1977) was considered most appropriate for examining the impact of broadband on household consumers. This was because this relationship was proposed for the context of household technology and time change, and was successfully applied to investigate the impact of computer use in the household (Vitalari et al., 1985). Two underlying principles were adopted from Robinson (1977) which state that: (1) a natural system like a household tends to be at an equilibrium, and (2) any new change due to equilibrium disturbance caused by triggers such as the use of new technology is adjusted by similar but already existing factors in the system, thereby restoring the equilibrium condition (Robinson, 1977; Vitalari et al., 1985). These two principles were utilised to propose the homeostatic model of the effect of computer use on household time allocation patterns (Vitalari et al., 1985). This model suggests three possibilities that may occur after introducing new technology in the household. These three possibilities are: absence of change in the existing patterns of household behaviour due to the non-use of the computer; short-term perturbations in household behaviour, and changes that signal long-term impacts (Vitalari et al., 1985). This model also suggests that consumer activities that are affected the most are internal. Examples of internal activities include watching television and hobbies, as opposed to external activities such as sports, eating, socialising with friends (Vitalari et al., 1985). This is a brief account of the homeostatic model; however, for a detailed discussion the reader may refer to the original source published by Vitalari et al. (1985).

The model of homeostasis discussed was utilised to conceptualise the impact of broadband on the time allocation patterns of households. Similar to the previous study (Vitalari et al., 1985), the model begins with an equilibrium state, where households do not have an Internet connection at home. Subscribing to the Internet at home may affect a consumer in one of the following three ways. First, there is a lack of change in the existing pattern of behaviour. This is because the characteristics of household users, such as age, are such that they elicit a low level of interest in the Internet. Second, there are short-term changes in the behaviour of the household. Third, there are long-term changes in the household's behaviour. Both short-term and long-term changes can affect time allocation patterns on both cognate activities such as television watching and reading, and differentiated activities such as sports and outdoor recreation (Vitalari et al., 1985). This research postulates that the magnitude of both short-term and long-term changes are dependent upon the type of Internet connection, i.e., narrowband or broadband, rate of use, and the length of Internet subscription in the household.

It is also assumed that the decrease in online shopping or the decrease in telephone conversations require time to develop trust and become habitual; therefore, the length of Internet subscription is likely to affect this for the long-term. Activities such as working at home, working in the office and commuting in traffic are interconnected and are more likely to be affected by the type of Internet connection. Households with broadband connections are more likely to have individuals who work at home in comparison to narrowband consumers. It can be argued that short-term changes may become long-term if Internet use is continued over time. For example, online shopping may be an initial instance of a short-term change in consumer behaviour; however, if there is continuous use of the Internet and a consumer develops trust and routine, then it may become a long-term change in household behaviour. Therefore, this research will answer the following research question:

RQ8: *Does the use of broadband affect the time spent on various daily life activities?*

With this question, the discussion on the theoretical aspects surrounding this research is drawn to a close. The next section summarises and concludes this chapter.

Summary

This chapter reviewed the various technology adoption and diffusion related theories and models including the DI, TRA, TPB, DTPB, TAM, MATH, and UD. The analysis suggests that although none of these theories and models could be as such applied to examine the broadband adoption, usage and impact, integrating the constructs across the models will be more appropriate and will assist in providing a coherent understanding of the research problem. Therefore, the most appropriate theories and models such as MATH, the TPB, DTPB and DI have been considered to be guiding frameworks for current research.

Second, this chapter has identified the factors that are expected to predict the BI to adopt broadband, which ultimately explains the broadband adoption behaviour. Also, the broadband adoption behaviour is expected to differentiate between the rate and variety of Internet use between the broadband and narrowband users. Using these factors, a conceptual model of broadband diffusion was developed. The proposed conceptual model is based on the assumption that the attitudinal, normative and control factors listed in Table 2.1 are responsible for influencing the BI to adopt broadband, which in turn is expected to predict broadband adoption behaviour. The proposed model also includes constructs to investigate whether broadband users differ from narrowband users when determining the usage and impact of the Internet. Whilst discussing the aforementioned factors, the underlying hypotheses (Table 2.1) were also proposed that need to be tested in order to verify the model. In order to test the underlying hypotheses that can verify the proposed conceptual model, the next step was to determine the relevant research method. Following that, it was essential to develop a reliable data collection instrument that could be utilised to collect empirical data from the household consumers.

In the following chapters, the proposed conceptual model will be utilised as a basis for empirical investigation. Chapter III begins with a discussion of the chosen research method. Following that, Chapters IV and V will provide a detailed discussion on the instrument development.

References

Ajzen, I. (1985). From intentions to actions: A theory of planned behaviour. In J. Kuhl & J. Beckmann (Eds.), *Action control: From cognition to behavior* (pp. 11-39). Heidelberg: Springer.

Ajzen, I. (1988). *Attitudes, personality, and behaviour*. Chicago: The Dorsey Press.

Ajzen, I. (1991). The theory of planned behaviour. *Organisational Behaviour and Human Decision Processes, 50,* 179-211.

Ajzen, I., & Fishbein, M. (1980). *Understanding attitudes and predicting social behavior*. NJ: Prentice-Hall.

Ajzen, I., & Madden, T. J. (1986). Prediction of goal directed behaviour: Attitudes, intentions, and perceived behavioural control. *J. of Exp. Social Psychology, 22,* 453-474.

Al-Jabri, I. M. (1996). Gender differences in computer attitudes among secondary school students in Saudi Arabia. *Journal of Computer Information Systems, 37*(1), 70-75.

Anckar, B. (2003). Drivers and inhibitors to E-commerce adoption: exploring the rationality of consumer behaviour in the electronic marketplace. In C. Ciborra et al. (Eds.), *Proceedings of the 11th ECIS on New Paradigms in Organizations, Markets and Society*, Napoli, Italy.

Anderson, B., Gale, C., Jones, M. L. R., & McWilliam, A. (2002). Domesticating broadband-what consumers really do with flat rate, always-on and fast Internet access? *BT Technology Journal, 20*(1), 103-114.

Anderson, B., McWilliam, A., Lacohee, H., Clucas, E., & Gershuny, J. (1999). Family life in digital home- domestic telecommunications at the end of the 20th century. *BT Technology Journal, 17*(1), 85-97.

Anderson, B., & Tracey, K. (2001). Digital living: The impact (or Otherwise) of the Internet on everyday life. *American Behavioral Scientist, 45,* 456-475.

Anderson, T. (2000). Regulation part 2: *Digital transactions area a cause for concern*. Retrieved September 15, 2004, from at http://www.netimperative.com/2000/05/04/Regulation_Part_2

Bagozzi, R. P. (1981). Attitudes, intentions and behaviour: A test of some key hypotheses. *Journal of Personality and Social Psychology, 41,* 607-627.

Benbasat, I., & Zmud, R. W. (1999). Empirical research in information systems: The practice of relevance. *MIS Quarterly, 23*, 3-17.

Brancheau, J. C., & Wetherbe, J. C. (1990). The adoption of spreadsheet software: Testing innovation diffusion theory in the context of end-user computing. *Information Systems Research, 1*, 115-143.

Brown, S., & Venkatesh, V. (2003). Bringing non-adopters along: The challenge facing the PC industry. *Communications of the ACM, 46*(4), 76-80.

BSG Briefing Paper. (2004). *The impact of broadband-enabled ICT, content, applications and services on the UK economy and society to 2010*. Retrieved November 15, 2004, from http://www.broadbanduk.org/news/news_pdfs/Sept%202004/BSG_Phase_2_BB_Impact_BackgroundPaper_Sept04(1).pdf

BSG Report. (2001). Report and strategic recommendations. Retrieved November 30, 2002, from http://www.broadbanduk.org/reports/BSG_Report1.pdf

Burgess, R. (1986). *Key variables in social investigation*. London: Routledge.

Burnkrant, R. E., & Page, T. J. (1988). The structure and antecedents of the normative and attitudinal components of Fishbein's theory of reasoned action. *Journal of Experimental Social Psychology, 24*, 66-87.

Carriere, R., Rose, J., Sirois, L., Turcotte, N., & Christian, Z. (2000). Broadband changes everything. Retrieved November 30, 2002, from http://www.mckinsey.de/_downloads/knowmatters/telecommunications/broadband_changes.pdf

Carveth, R., & Kretchmer, S. B. (2002). The digital divide in Western Europe: Problems and prospects. *Informing Science, 5*(3), 239-249.

Chau, P. Y. K., & Hu, P. J. H. (2001). Information technology acceptance. *Decision Sciences, 32*, 699-719.

Chen, Y., Lou, H., & Luo, W. (2001-2002). Distance learning technology adoption: A motivation perspective. *Journal of Computer Information Systems, 42*(2), 38-43.

Chinn, M. D., & Fairlie, R .W. (2004). The determinants of the global digital divide: A cross-country analysis of computer and Internet penetration. *Discussion Paper Series*, Institute for the Study of Labour, Bonn, Germany.

Choudrie, J., & Lee, H. (2004). Broadband development in South Korea: Institutional and cultural factor. *European Journal of Information Systems, 13*(2), 103-114.

Crabtree, J. (2003). *Fat pipes, connected people-rethinking broadband Britain*. Retrieved March 30, 2004, from iSOCIETY Report http://www.theworkfoundation.com/pdf/1843730146.pdf

Davis, F. D. (1989). Perceived usefulness, perceived ease of use, and user acceptance of information technology. *MIS Quarterly, 13*, 319-340.

Davis, F. D., Bagozzi, R. P., & Warshaw, P. R. (1989). User acceptance of computer technology: a comparison of two theoretical models. *Management Science, 35*(8), 982-1003.

DeLone, W. H., & McLean, E. R. (1992). Information systems success: The quest for the dependent variable. *Information Systems Research, 3*(1), 60-95.

DeLone, W. H., & McLean, E. R. (2003). The DeLone and McLean model of information systems success: A ten-year update. *J. of Mgt. Information Systems, 19*(4), 9-30.

DeSanctis, G., & Poole, M. S. (1994). Capturing the complexity in advanced technology use: Adaptive structuration theory. *Organization Science, 5,* 121-147.

Dwivedi, Y. K., Choudrie, J., & Gopal, U. (2003). Broadband stakeholders analysis: ISPs perspective. In R. Cooper et al. (Eds.), *Proceedings of the ITS Asia-Australasian Regional Conference*, Perth, Australia.

Finch, J. (1986). Age. In R. Burgess (Ed.), *Key variables in social investigation*. London: Routledge.

Fishbein, M., & Ajzen, I. (1975). *Belief, attitude, intention, and behavior: An introduction to theory and research*. Reading, MA: Addison-Wesley.

Foxall, G. R. (1992). The behavioural perspective model of purchase and consumption: From consumer theory to marketing practice. *Journal of the Academy of Marketing Science, 20*, 189-198.

Fulk, J., & Boyd, B. (1991). Emerging theories of communication in organisations. *Journal of Management, 17*(2), 407-446.

Fulk, J., Steinfield, C. W., Schmitz, J., & Power, J. G. (1987). A social information processing model of media use in organisations. *Communication Research, 14*(5), 529-552.

Gefen, D., & Straub, D. W. (1997). Gender differences in the perception and use of e-mail: An extension to the technology acceptance model. *MIS Quarterly, 21*(4), 389-401.

Gilligan, C., & Wilson, R. M. S. (2003). *Strategic marketing planning*. Oxford: Butterworth-Heinemann.

Harris, A. L., Medlin, D., & Dave, D. S. (1996). Multimedia technology as a learning tool: A study of demographic and cultural impacts. *Journal of Computer Information Systems, 36*(4), 18-21.

Heijden, H. (2004). User acceptance of hedonic information systems. *MIS Quarterly, 28*(4), 695-705.

Hoffer, J. A., & Alexander, M. B. (1992). The diffusion of database machines. *Database, 23*, 13-20.

Holmes, M. C. (1997). Comparison of gender differences among information systems professionals: A cultural perspective. *Journal of Computer Information Systems, 38*(4), 78-86.

Hu, P. J., Chau, P. Y. K., Sheng, O. R. L., & Tam, Y. K. (1999). Examining the technology acceptance model using physician acceptance of telemedicine technology. *Journal of Management Information Systems, 16*, 91-112.

Huang, W., Ambra, J. D., & Bhalla, V. (2002). An empirical investigation of the adoption of e-Government in Australian citizens: Some unexpected research findings. *Journal of Computer Information Systems, 43*(1), 15-22.

Jackson, S., & Scott, S. (2001). *Gender*. London: Routledge.

Lacohee, H., & Anderson, B. (2001). Interacting with telephone. *International Journal of Human-Computer Studies, 54*, 665-699.

Lee, H., O'Keefe, B., & Yun, K. (2003). The growth of broadband and electronic commerce in South Korea: Contributing factors. *The Information Society, 19*, 81-93.

Lee, H., & Choudrie, J. (2002). *Investigating broadband technology deployment in South Korea*. Brunel- DTI International Technology Services Mission to South Korea. DISC, Brunel University, Uxbridge, UK.

Mason, S. M., & Hacker, K. L. (2003). Applying communication theory to digital divide research. *IT & Society, 1*(5), 40-55.

Mathieson, K. (1991). Predicting user intentions: Comparing the technology acceptance model with the theory of planned behaviour. *Infor. Systems Research, 2*(3), 173-191.

McCalla, R., & Ezingeard, J. N. (2005, June 26-28). Examining the link between technology use, emotional expression and service quality perceptions: The data collection protocol. In *Proc. of the 13th European Conf. on Information Systems*, Regensberg, Germany.

Moore, G. C. (1987). End-user computing and office automation: A diffusion of innovation perspective. *Infor, 25*, 214-235.

Moore, G. C., & Benbasat, I. (1991). Development of an instrument to measure the perceptions of adopting an information technology innovation. *Information Systems Research, 2*(3), 192-222.

Morgan, D. H. J. (1986). Gender. In R. Burgess (Ed.), *Key variables in social investigation*. London: Routledge.

Morris, M. G., & Venkatesh V. (2000). Age differences in technology adoption decisions: Implications for a changing work force. *Personnel Psychology, 53*(2), 375-403

Nilikanta, S., & Scammel, W. (1990). The effect of information sources and communication channels on the diffusion of innovation in a database development environment. *Management Science, 36*, 24-40.

Oh, S., Ahn, J., & Kim, B. (2003). Adoption of broadband Internet in Korea: The role of experience in building attitude. *Journal of Information Technology, 18*(4), 267-280.

Oliver, R. L., & Bearden, W. O. (1985). Crossover effects in the theory of reasoned action: A moderating influence attempt. *Journal of Consumer Research, 12,* 324-340.

Parasuraman, A., Berry, L., & Zeithaml, V. A. (1991). Refinement and assessment of the 'SERVQUAL' Scale. *Journal of Retailing, 67*(4), 420-451.

Parasuraman, A., Zeithaml, V. A., & Malhotra, A. (2005). E-S-QUAL: A multiple-item scale for assessing electronic service quality. *Journal of Service Research, 7*(3), 213-234.

Pearson, M. J., Crosby, L., Bahmanziari, T., & Conrad, E. (2002). An empirical investigation into the relationship between organisational culture and computer efficacy as moderated by age and gender. *Journal of Computer Information Systems, 43*(2), 58-70.

Rice, C. (1997). *Understanding customers.* Oxford: Butter worth-Heinemann.

Rice, R. E., Grant, A. E., Schmitz, J., & Torobin, J. (1990). Individual and network influences on the adoption and perceived outcomes of electronic messaging. *Social Networks, 12*(1), 27-55.

Robinson, J. P. (1977). *How Americans use time.* New York: Praeger.

Rogers, E. M. (1995). *Diffusion of innovations.* New York: Free Press.

Rosemann, M., & Vessey, I. (2005, June 26-28). Linking theory and practice: Performing a reality check on a model of IS success. In *Proceedings of the 13th European Conference on Information Systems*, Regensberg, Germany.

Salancik, G. R., & Pfeffer, J. (1978). A social information processing approach to job attitudes and task design. *Administrative Science Quarterly, 23*(2), 224-253.

Sawyer, S., Allen, J. P., & Heejin, L. (2003). Broadband and mobile opportunities: A socio-technical perspective. *Journal of Information Technology, 18*(4), 121-136.

Schifter, D. B., & Ajzen, I. (1985). Intention, perceived control, and weight loss: An application of the theory of planned behavior. *J. of Personality and Social Psych., 49*, 843-851.

Shih, C. F., & Venkatesh, A. (2004). Beyond adoption: Development and application of a use-diffusion model. *Journal of Marketing, 68*, 59-72.

Shih, C. F., & Venkatesh, A. (2003). A comparative study of home computer adoption and use in three countries: U.S., SWEDEN, and INDIA, CRITO Working

Paper, Retrieved July 20, 2004, from http://www.crito.uci.edu/noah/publications.htm

Shimp, T. A., & Kavas, A. (1984). The theory of reasoned action applied to coupon usage. *Journal of Consumer Research, 11*, 795-809.

Solomon, S. L., & O'Brien, J. A. (1990). The effect of demographic factors on attitudes toward software piracy. *Journal of Computer Information Systems, 30*(3), 40-46.

Stanton, L. J. (2004). Factors influencing the adoption of residential broadband connections to Internet. In *Proceedings of the 37th Hawaii Int. Conference on System Sciences.*

Tan, M., & Teo, T. S. H. (2000). Factors influencing the adoption of Internet banking. *Journal of the Association for the Information Systems, 1.*

Taylor, S., & Todd, P. A. (1995). Understanding information technology usage: A test of competing models. *Information Systems Research, 6*(1), 44-176.

Tornatzky, L. G., & Klein, K. J. (1982). Innovation characteristics and innovation adoption-implementation: A meta-analysis of findings. *IEEE Transactions on Engineering Management, 29*, 28-45.

Venkatesh, V. & Brown, S. (2001). A longitudinal investigation of personal computers in homes: Adoption determinants and emerging challenges. *MIS Quarterly, 25*(1), 71-102.

Venkatesh, V. & Morris, M. G. (2000). Why don't men ever stop to ask for directions? Gender, social influence, and their role in technology acceptance and usage behaviour. *MIS Quarterly, 24*(1), 115-139.

Venkatesh, V., Morris, M. G., Davis, B. G., & Davis, F. D. (2003). User acceptance of information technology: Toward a unified view. *MIS Quarterly, 27*(3), 425-478.

Venkatesh, A., Shih, C. F. E., & Stolzoff, N. C. (2000). A longitudinal analysis of computing in the home census data 1984-1997. In A. Sloane & F. van Rijn (Eds.), *Home informatics and telematics: Information, technology and society* (pp. 205-215). Norwell, MA: Kluwer Academic Publisher.

Vitalari, N. P., Venkatesh, A., & Gronhaug, K. (1985). Computing in the home: shifts in the time allocation patterns of households. *Comm. of the ACM, 28*(5), 512-522.

Wood, W., & Glass, R. (1995-1996). Sex as a determinant of software piracy. *Journal of Computer Information Systems, 36*(2), 37-40.

Yang, Z., Cai, S., Zhou, Z., & Zhou, N. (2005). Development and validation of an instrument to measure user perceived quality of information presenting Web portals. *Information & Management, 42*(4), 575.

Section 1.2

The Methodological Underpinning

Chapter 3

Research Methodology

Abstract

The previous chapter formed a conceptual model aimed at examining broadband adoption, usage, and impact from the context of household consumers. This chapter aims to provide an overview of the research approaches utilised within information systems (IS) field, which leads to the selection of an appropriate research approach for guiding the validation of the conceptual model. To understand the research topic, to validate and understand the conceptual model, and to obtain the required data, a quantitative research was employed. The philosophical foundation utilised for guidance was positivism, resulting in a survey research approach being employed. The data collection technique used to collect the data was the questionnaire. Reasons for the selection of the philosophical underpinning, type of research approach, and data collection method are explained and justified within this chapter. This chapter is structured as follows. The next section provides an overview of the underlying epistemologies and then provides justifications for the preferred one. This is followed by an overview discussion on various issues related to the available research approaches in the IS field and a justification for the selection of a survey as the research approach. A detailed account of the various aspects of the survey approach is then offered. This is followed by a brief discussion on issues relating to data analyses. The final section offers a summary of the chapter.

Underlying Epistemology

According to Myers (1997), epistemology refers to the assumptions about knowledge and how it can be obtained. Within IS, there are three underlying epistemologies that researchers can select in order to guide a particular research. These are positivism, interpretivism, and critical research (Mingers, 2001, 2003; Orlikowski & Baroudi, 1991).

A large variation is reported in terms of the percentage use of these underlying epistemologies within IS research. Orlikowski and Baroudi's (1991) study found that amongst the three IS epistemologies, positivism emerged in 96.8 percent of studies. In contrast, only 3.2 percent of studies employed interpretive epistemology and at the extreme end, none of the studies could be placed within critical epistemology (Orlikowski & Baroudi, 1991). However, more than a decade has passed since the Orlikowski and Baroudi (1991) study and the percentage use of underlying epistemologies within IS research has slightly changed. A Mingers (2003) study suggests that 75 percent of the IS research employs a positivist approach, 17 percent interpretivist and only 5 percent critical research (Mingers, 2003). The statistics suggest that positivism is the most favoured underlying epistemology within IS research.

Orlikowski and Baroudi (1991) defined research as positivist if there was evidence of formal propositions, quantifiable measures of variables, hypothesis testing, and the drawing of inferences about a phenomenon from the sample to a stated population (Orlikowski & Baroudi, 1991). Straub et al., (2005) described positivism from the statistical point of view by suggesting that the objective of statistics (mainly T, F, and Chi-square statistics) employed by the quantitative positivist research (QPR) is to falsify the null hypothesis, which is the assumption that the data in the dependent variable are not affected by the data in the independent variable(s). Since each theoretical hypothesis (the hypothesis as stated in the theory) should be the exact opposite of its null hypothesis by predicting a difference in the dependent variable, it follows logically that if the null hypothesis is rejected, then presumably the theoretical hypothesis is supported (Straub et al., 2005).

Since this research provides evidence of propositions (Chapter 2), quantifiable measures of variables (Chapters 4 and 5), hypothesis testing and the drawing of inferences about a phenomenon from the sample to a stated population (Chapters 6, 7, and 8), the positivist epistemology was considered to be appropriate for this research. In contrast, since interpretivist epistemology focuses upon the complexity of human sense making, it was necessary to pursue research employing qualitative data collection (Kaplan & Maxwell, 1994; Myers, 1997; Straub et al., 2005; Walsham, 1993) with very limited respondents.

Thus, interpretivist epistemology was considered to be less relevant for this research. Furthermore, the critical epistemology was also considered to be less appropriate for undertaking this research. This is because the purpose of the current research is not to focus upon the oppositions, conflicts, and contradictions (Myers, 1997). Instead it investigates factors that at this particular point of time are affecting the adoption of broadband. After considering all the three underlying epistemologies, this research adopted positivist epistemology. This is because technology adoption and diffusion is considered to be one of most mature areas within IS research. Due to its long tradition of research, a number of theories and models have been developed and validated for examining a variety of technological objects. Consequently, a variety of constructs (dependent and independent variables) suitable for diverse situations are available, which can rationally be adapted to examine the adoption and diffusion of new technologies (Venkatesh et al., 2003). This was the basis for developing a conceptual model of broadband adoption and formulating the research hypotheses presented in Chapter 2.

Following Straub et al. (2005) description of positivism, this research will employ statistics, such as the T, F, and Chi-square test, to determine if this data supports the research hypotheses (Straub et al., 2005). This research does not suggest that the other two epistemologies cannot be applied to this research. It is argued that for this research context, positivism is much more appropriate and feasible.

Quantitative and Qualitative Data

According to Myers (1997), qualitative data is derived from various sources that include observation and participant observation (fieldwork), interviews and questionnaires, documents and texts, and the researcher's impressions and reactions. Qualitative data are useful means to understand people and the social and cultural contexts within which they live (Myers, 1997). Cornford and Smithson (1996) describe quantitative data as metrics (numbers) that can be used to describe the phenomenon (objects and relationships) under study. Straub et al. (2005) argued that the numbers come to represent values and levels of theoretical constructs and concepts and the interpretation of the numbers is viewed as strong scientific evidence of how a phenomenon works (Straub et al., 2005). Sources of quantitative data in the social sciences include survey methods, laboratory experiments, formal methods (e.g. econometrics) and numerical methods such as mathematical modeling (Myers, 1997, Straub et al., 2005).

Since data utilised in this research were collected employing survey methods (Myers, 1997) and represent values and levels of theoretical constructs (Myers, 1997; Straub et al., 2005) such as relative advantage, utilitarian outcomes, hedonic outcomes, service quality, primary influence, secondary influence, self-efficacy, knowledge, facilitating conditions resources, and behavioural intentions, the data collected in this research belongs to the quantitative category rather than the qualitative.

Research Approaches

Taxonomy of IS Research Approaches

When conducting any research, selecting an appropriate method is a critical issue. In the IS area, several attempts have been made to review and classify research approaches (Galliers, 1992; Galliers & Land, 1987; Mingers, 2001; 2003; Nandha-kumar & Jones, 1997; Orlikowski & Baroudi, 1991; Walsham, 1995a, 1995b). Early work was undertaken by Galliers (1992), who provided a taxonomy of prevalent IS research approaches. This taxonomy considered a range of *positivist* and *interpretive* research approaches including experiments, surveys, case studies, theorem proof, forecasting, simulation, reviews, action research, futures research, and role/game playing (as shown in *Appendix 3.1*).

The other early research was by Orlikowski and Baroudi (1991), who offered a philosophically reflective paper with a North American perspective. In this work, the emphasis was on categorising published IS research according to the used epistemolo-gies, and it was found that although positivism was prevalent, *critical epistemology* (Orlikowski & Baroudi, 1991) was also beginning to emerge. Similar to Orlikowski and Baroudi (1991), a recent classification by Minger (2003) also categorised IS research approaches into three categories, including critical research. Straub et al., (2005) have divided positivist research in two categories, namely quantitative positivist research (QPR), such as lab and field experiment, field study, and non-quantitative positivist research (non-QPR), for example case study and participative research. A list of the IS research approaches is offered in *Appendix 3.1*.

The extant IS literature suggests that different terms are used for the same research methods (Mingers, 2003). For instance, the terms 'survey' and 'questionnaire' are used indistinguishably (Mingers, 2003). Contrastingly, the terms 'case study' and 'interviews' are used synonymously, although they are distinct from each other (Mingers, 2003). Bearing this in mind, it was felt necessary to clarify the various terms that are used for different types of research approaches. For this purpose, Mingers' classification and description of research approaches (Mingers, 2003) was adopted for this research. This classification was followed for two reasons: first, Mingers' research is the most recently published work; second, it encompasses a variety of research approaches associated with all three epistemological standpoints. As a reminder, positivist research methods include observations, measurements, surveys, questionnaires, instruments, laboratory and field experiments, statistical analysis, simulations, and case studies. Interpretivist research methods consist of interviews, qualitative content analysis, ethnography, grounded theory, and participant observa-tion. Finally, the critical standpoint involves intervention and change, employing the methods of action research, critical theory, and consultancy (Mingers, 2003).

Trend of Research Approaches Used in Information Systems

Mingers (2001, 2003) conducted a review of all the papers published during 1993-1998, in two leading American journals (*MIS Quarterly (MISQ)* and *Information Systems Research (ISR)*) and four European ones (*European Journal of Information Systems (EJIS), Information Systems Journal (ISJ), Journal of Accounting, Management and IT (JAMIT),* and *Journal of Information Technology (JIT)*). The findings of this study suggest that about 80 percent of the evaluated papers contain some form of empirical research, where surveys, interviews, experiments and case studies are the dominant approaches. Alternatively, approaches like participant observation, grounded theory and soft systems methodology are rarely used.

These studies provide evidence that although several research methods are suggested (Galliers, 1992; Minger, 2003; Orlikowski & Baroudi, 1991), only surveys, experiments, interviews, and case studies are predominantly used within the IS area.

Trend of Research Approaches Used within Technology Adoption Research

Previous studies have focused upon examining the trends of IS research approaches in general. However, little information is available for the trends of the research approaches used for the specific research domain, such as technology adoption and diffusion. Therefore, an examination of the literature was conducted to ascertain the research approaches employed in the area of technology adoption and diffusion. This was pursued by examining the following:

1. Since focus of this research is broadband adoption and use, prevalence of different research approaches in the area of technology adoption were examined.

2. Since the context of this research is the household, the prevalence of different research approaches in the area of technology adoption and use within the household context were examined.

3. In order to determine an appropriate research approach for examining household consumers, the relationship between the research approaches used and the types of *unit of analysis* (i.e., users, consumers, organisations) was examined.

To explore the trends of IS research approaches, a review of articles was undertaken, selecting from those published within peer-reviewed and highly rated journals including *MISQ, EJIS, ISJ, ISR*, and other relevant publications. Since Mingers' (2003) method was used, the sample selection used in this research was very close to this study. Since these journals were examined in Mingers' (2003) study, they were also selected for this study. To avoid any bias and to obtain a common perspective, two American and two European IS Journals were examined to review trends of research approaches in the technology adoption and usage area.

A total of 633 published articles appearing during the 1992-2003 period in four IS journals were examined to select the empirical papers addressing the issue of technology adoption and usage. This research followed Mingers' definition of an empirical paper, which states that a paper is regarded as empirical if it reports on new data (of any kind) that has been generated by the underlying research and where the resultant analysis is a substantive part of the paper's contribution (Mingers, 2003). Empirical papers focusing on the area of technology adoption were then studied and their research approach was recorded. Since IS research on technology adoption focuses on *users* as research artefacts, another wider search of the relevant literature was conducted. This was done in order to examine the approaches used to study adoption and usage of technology in the *household* context.

From the 633 examined articles, 31 articles (4.9 percent) addressed issues related to technology adoption. This proportion in specific IS journals were as follows: *MISQ* (6 percent), *ISR* (5.15 percent), *EJIS* (5 percent) and *ISJ* (2.63 percent).

The analysis of the articles suggests that the researchers investigating technology adoption used two main research approaches: the *survey* and *case study*. 74 percent of the articles employed the survey approach, which suggests that it is the most widely used approach in technology adoption research. This then led this research to consider the survey approach.

Further support was obtained from previous findings and evidence that the survey approach is more dominant in the IS area (Farhoomand, 1992; Mingers, 2001, 2003; Orlikowski & Baroudi, 1991). The remaining 26 percent of the research employed the case study method. No other approaches were employed to investigate the use or adoption of technology. Another interesting observation is that the case study approach was exclusively employed to study the organisational adoption of technology, while surveys were used to study a range of contexts. For example, surveys were used to study technology adoption within the context of technology users, household and online consumers, senior executives, and small firms. Although technology adoption is a common topic within the IS area (Venkatesh et al., 2003), the research approaches used are of very limited diversity.

The review of the previous articles also suggests that research on technology adoption in the context of the household has just begun to emerge. IS researchers have mainly focused on organisational issues. Therefore, another attempt was made to identify publications, which addressed technology adoption issues in the household context. For this purpose, articles were extracted from both the IS and non-IS journals including *Advances in Consumer Research, American Behavioral Scientist, Journal of Marketing, Management Science,* and *Journal of Economic Psychology.*

Analysis of the selected articles indicated that the survey approach was once again dominant in the study of consumer adoption of technology in the household context. The survey approach was employed in 63 percent of the articles. The range of tools employed to conduct a survey include the postal service, telephone, face-to-face interviews and web questionnaires. Twenty-five percent of the research reported was conducted using a multi-method approach, whereby a combination of the survey with either an interview or time-use diary was employed. Other methods used were the ethnographic study and analysis of secondary data obtained from census figures. For the purpose of investigating the *adoption* of technology (especially ICTs) within the household, the survey approach seemed to be the most predominant. Other approaches including the multi-method, ethnographic study, and secondary data analysis were employed mostly for investigating the *usage* of technology in the household.

The findings suggest that the survey was the most widely used approach to examine the technology adoption issues both in the context of the organisations and households. The case study approach was employed only for investigating technology adoption issues in the organisational context, particularly when the *unit of analysis* was the organisation rather than the individual users. This approach was not employed in the household context. Other approaches such as the ethnographic study and time-use diaries were employed in the household context, but not in the organisational context.

Justification for Survey as a Preferred Research Approach

From the findings highlighted in the previous section, it can be concluded that although a range of research approaches are available to IS researchers, the survey as a research approach is most widely employed for examining technology adoption related issues. The choice of approach seems to correspond with the unit of analysis. Where researchers considered the organisation as a unit of analysis, the case study approach was favoured. In studies relating to individual users or consumers, the survey approach was favoured. This can be attributed to issues such as convenience, cost, time and accessibility (Gilbert, 2001). The extent to which a researcher can be a part of the context being studied is also a factor that plays an important role in determining a research approach. Within the household context, it is difficult for a researcher to be a part of the context; therefore the survey approach would be more feasible than others, such as ethnography and observations. Furthermore, the aim of this research was to examine broadband adoption and diffusion across the UK or using a nationwide perspective. Hence, in order to get an overall picture of the research issue, collecting data from a large number of participants from across the UK was required. This means employing any other approach such as ethnography that utilises an interview or observation, as data collection tools demand huge amounts of financial resources, manpower, and time. As this is a student research project, all three factors limited the ability of the researcher when investigating this research issue.

Selection of the approach in this case was also influenced by the type of theory and models employed to examine broadband adoption and diffusion research (Chapter 2). The conceptual model proposed in Chapter 2 includes a number of research hypotheses that need to be tested before concluding this study. This requires collecting quantitative data and statistical analysis in order to test the research hypotheses. Although a number of research approaches are available within the category of quantitative positivist research (Straub et al., 2005, Appendix 3.1), the survey is the only appropriate research approach that can be employed to conduct such research (i.e., that requires hypotheses testing and validation of the conceptual model) in a social setting, in this instance the household. One of the planned contributions of this study is to provide insights to the Internet service providers (ISPs) about the factors that are salient to consumer adoption and non-adoption of broadband and to establish a relationship with behavioural intention and actual adoption.

In order to achieve this, it was essential to collect quantitative data on a number of variables including demographics and thereafter perform a regression analysis to identify a relationship. This was again a logical reason for adopting the survey as a research approach and collect quantitative data that may help ISPs to understand the behaviour of household consumers and their demographic characteristics, in order to encourage and promote broadband adoption.

On the basis of the stated reasoning, it was decided that the survey is one of the most appropriate and feasible research approaches to conduct this research. The next section provides details on the strategy that was used to execute this research, followed by a detailed discussion on various aspects of the survey approach in the context of broadband adoption, usage and impact related issues.

Survey Research Approach

Before proceeding into details about the various aspects of the survey, it is important to clarify the term 'survey.' The majority of IS research approaches' classifications (Galliers, 1992; Galliers & Land, 1987; Mingers, 2001, 2003; Nandhakumar & Jones, 1997; Orlikowski & Baroudi, 1991) have employed the term 'survey' as an approach within which a number of data collection techniques, such as mail, telephone, and interviews are available and can be utilised for data collection purposes. However, Straub et al. (2005) have denoted the term 'survey' as a data collection technique along with others such as interviews within the field study as a research approach. This study adopts the first view that denotes the survey as a research approach (Galliers,

1992; Galliers & Land, 1987; Mingers, 2001,2003; Nandhakumar & Jones, 1997; Orlikowski & Baroudi, 1991). This is because the survey as a research approach is the most widely accepted view and is older in its acceptance within the research community. Therefore, within this study the term 'survey' represents a research approach that is utilised to conduct this study. The data collection technique in this research is also referred to as a questionnaire or research instrument. To provide more information on the survey, the following subsections discuss various aspects of the survey research approach.

There are three essential components of the survey research approach, which are sampling, data collection, and instrument development (Fowler, 2002). Fowler (2002) suggested that it is obligatory for a good survey design to combine all these components. The first component, sampling, involves the selection of a small subset of a population that is representative of the whole population. The most important thing to consider in a good sample is applying a technique that gives all or nearly all the population members the same chance of being selected (Fowler, 2002). The data collection can be conducted employing techniques, such as in-person, telephone, mail, and the Internet; however, the selection should be made after evaluating the advantages and disadvantages from the perspective of a particular research context (Fowler, 2002). Therefore, in order to evaluate the advantages and disadvantages from the perspective of this research, various aspects of sampling and data collection will be discussed. The third component, 'instrument development,' is briefly introduced in the following section and a detailed description is provided in the next chapter (Chapter 4).

Sampling

Fowler (2002) has suggested following five critical issues with regards to sampling. These issues consist of: (1) the choice of whether or not to use a probability sample; (2) the sample frame (those people who actually have the opportunity to be sampled); (3) the size of the sample; (4) the sample design (the particular strategy used for sampling a household consumer); and (5) the rate of response (the percentage of those sampled from whom the data are actually collected). The first four issues are discussed within this section and the remaining issue, response rate, will be discussed in the successive section.

The Sample Frame

According to Fowler (2002), in a sample selection procedure, people who have a chance of being included among those being selected constitute the sample frame, and that is considered to be a primary step towards evaluating the quality of a sample. Three ways a sample selection can be made are as follows (Fowler, 2002): (1) from a more or less complete list of individuals in the studied population; (2) from a set of people who go somewhere or do something that enables them to be sampled; and (3) as in this method, sampling is done in two or more stages: the first stage involves creating a list of individuals and the second sample selection step is made from the created list.

Three characteristics comprising comprehensiveness, probability of selection, and efficiency of the sample frame need to be evaluated in order to select an appropriate sample frame (Fowler, 2002). If any sample frame excludes some proportion of the selected population then it is considered to be less comprehensive. Such a sample frame is less preferred when considering the selection of sample frame (Fowler, 2002). Probability of selection is where each individual should appear once and only once in a good sample frame in order to provide an equal chance for every entry in the selection process. It means that the sample frame should be checked for repetitions of entries. If some entries are stated more than once in a sample frame then they have more chances to be selected than others. Therefore, such sample frames should be avoided or less preferred. When a sample frame does not include units that are not among those that the researcher wants to sample, then it is considered to be an efficient sampling frame (Fowler, 2002).

Rice (1997) suggested six criteria of a good sampling frame comprising completeness, accuracy, adequacy, being up-to-date, convenience, and non-duplication. *Completeness* refers to all the members of the population included in a list who have reliable addresses and information (i.e., *accuracy*). *Adequacy* is similar to completeness and refers to a sampling frame covering the entire population. *Up-to-date* simply refers to information included in the sampling frame that is regularly updated. *Convenience* refers to a sampling frame that is readily accessible. Finally, *non-duplication* refers to each member of the population appearing on a list only once (Rice, 1997). A recommended sample frame that satisfies all but one of the specified criteria for the UK population is the *electoral register* (a list of individuals) (Rice, 1997). This sample frame is readily available but, due to recent changes in the Data Privacy Act, the UK government prohibits access and use of the electoral role for research and marketing purposes.

In an e-mail communication with the British Library, it was suggested that an alternative to the electoral register is a commercially available CD Rom, called 'UK-Info Disk V11,' that consists of a degree of similar comprehensiveness, accuracy, adequacy, and up-to-date and non-duplicated information. The added advantages of this sample frame are that it is legally accessible/available for purchase and its mailing list can be prepared faster as there is no need to type addresses; instead this can be achieved just by copying and pasting relevant records. Therefore, due to the inaccessibility of the electoral register, it was decided that it is appropriate to utilise an alternative sample frame such as the 'UK-Info Disk V11' for the UK population. The UK-Info Disk V11 contained 31 million electoral register records, i.e., addresses of UK citizens.

Sampling Techniques

After determining that the UK-Info Disk V11 was a cost-effective, comprehensive, and efficient sample frame, the next step was to decide upon a selection technique for respondents to be included in the final study. Fowler (2002) suggests a number of techniques that can be utilised for selecting respondents from a sample frame. Amongst them the probability sampling techniques include simple random, systematic, and stratified sampling. This research utilised stratified random sampling as the sampling technique. A brief account on simple random, systematic and stratified sampling is provided and then the reasons for the selection of stratified sampling are now discussed.

Conducting a simple random sampling required a numbered list of the target population with each entry appearing once and only once. Then the required amount of random numbers needed to be generated within a specified range of numbers. This could be done using a computer program, a table of random numbers or some other generator of random numbers. Entries corresponding to the total amount of random numbers selected would then constitute a sample random sample of the target population (Fowler, 2002).

Since ordering and numbering a large target population can be cumbersome, laborious, and time consuming, an alternative to replace this technique is systematic sampling. Systematic sampling is not only mechanically easier to create but it also allows obtaining the benefits of stratification more easily without compromising the precision of sampling. Creating a systematic sample involves determining the total number of entries in a sample frame and then selecting a number of entries from the list. A division of latter values by the former one will produce a fraction. This estimated fraction is utilised as an interval every time a number is drawn and the composite of which makes the systematic sample (Fowler, 2002).

When a sample frame is divided by a number of subgroups on the basis of the characteristics of the target population and the total number of entries differing in the subgroups, then it is not considered appropriate to apply either simple random sampling or systematic sampling. For example, the UK population is divided into various boroughs that differ in terms of their total population; therefore, it is not appropriate to apply both simple random and systematic sampling. This is due to the difference in the size of the subgroups and the fact that entries from the larger subgroups would have more chances to be selected than the smaller ones. In this situation it is appropriate to apply the stratified random sampling techniques. The initial step of this sampling technique involves estimating how many entries need to be selected from each subgroup according to its total size. This can be achieved by dividing the total number of entries in a subgroup from the total sample size. Once the numbers of entries for all the estimated subgroups are obtained, the selection can be made from each subgroup according to the sample random or systematic sampling process. Thereafter, combining all the entries selected from the various subgroups offers a stratified random sample (Fowler, 2002).

Of the three sampling techniques, the structure of the sample frame (i.e., UK-Info Disk V11) of this research necessitated the adoption of the third approach- stratified random sampling. This is because within the UK-Info Disk V11 database, the entire UK population was alphabetically listed by family name. The number of entries varied according to the alphabet. Therefore, to obtain a sample in equal proportion from the entire alphabet, it was most suitable to adopt the stratified sampling approach. Initially, in order to draw a stratified random sample a respondent address from each letter of the alphabet was extracted. Then the sample size for each alphabet was determined according to the total sample size. Thereafter, a unique random number for each alphabetical letter was generated using the research randomiser software. Respondents corresponding to the random numbers were then selected for data collection from the sample frame (i.e., UK-Info Disk V11).

Sample Size

This is a most commonly encountered issue in survey research where researchers look for a basis to determine the sample size. There are three commonly used approaches to determine the sample size, but Fowler (2002) suggested that they are not an appropriate way to determine the sample size. These three commonly used approaches include determining the sample size based on the total size of the target population, deciding it upon the basis of a recommended standard size, and determining it based on how large a margin of error can be tolerated in a research.

The first approach of determining a sample is based on an inappropriate reasoning that the adequacy of a sample depends heavily on the fraction of the population included in that sample; in other words, a large fraction will make a sample more credible (Fowler, 2002). However, since the fraction of the population included in the sample is not a component of sampling error estimation (i.e., a measure of the precision of the sample), this approach is not an appropriate basis to determine the sample size (Fowler, 2002).

The second frequently used approach for determining a sample size is to derive it on the basis of the sample size in existing studies; which is inappropriately termed as a standard survey study (Fowler, 2002). As Fowler (2002) suggests, although it is feasible to consider the sample sizes of a particular population that competent studies have considered appropriate, the sample size decision must be determined on a case-by-case basis. The consideration should be based on the variety of goals to be achieved by a particular study and other related aspects of research design (Fowler, 2002).

The third approach of inappropriate determination of the sample size is based upon the consideration of the margin of error that is acceptable in a particular research, or the amount of precision expected from estimates (Fowler, 2002). Fowler (2002) argued that in theory there is nothing wrong with this approach; however, practically this approach provides little help to researchers in sample determination due to the following reason. The majority of survey studies involve several estimations and the desired precision for these estimates is likely to vary; therefore, it is not appropriate to make a sample size decision on the need for precision of a single estimate (Fowler, 2002). Furthermore, it is a less possible situation to specify an acceptable margin of error in advance.

This approach also assumes that errors only emerge from sampling and ignores the fact that there are several other sources of error such as response bias. Therefore, Fowler (2002) suggests that "the calculation of precision based solely on the sampling error is an unrealistic oversimplification, hence forming decisions based on the sample size is inappropriate" (Fowler, 2002).

According to Fowler (2002) the prerequisite for determining a sample size is a data analysis plan. Data analysis of the current study required utilising a number of statistical techniques, such as the principal component analysis (PCA), regression analysis, t-test and chi-square test. The following section provides a detailed discussion for using the statistical techniques in this research.

It has been suggested that in order to perform rigorous statistical analysis, such as the principal component analysis, the sample size should be above 300 (Stevens, 1996). Therefore, keeping the statistical analysis plan in mind, it was decided that the total sample size should be large enough to obtain a minimum of 300 responses. A rough estimate of the total sample size was determined by using the pilot response rate as a basis of the final survey.

Total sample size = [Total responses required X 100] / Pilot response rate

= 300 X 100 / 20 = **1500**

A sample size of 1500 was required to achieve 300 responses. To compensate for any shortfalls in the 300 responses that may occur due to any undelivered and partially completed responses, the sample size was further increased from 1500 to 1600. Therefore, a total sample size of 1600 was considered for this study.

Non-Response Bias and Response Rate

Non-Response

A non-response can be either of the following two types: (1) non-response to individual questions, i.e., not answering a few questions; and (2) not answering any questions or not returning the questionnaire at all. Occurrence of the first type of non-response is frequently low (Fowler, 2002) and does not contribute much towards an error. However, when its occurrence is high, it has the potential to affect a survey estimate. The reported occurrence of the second type of non-response is more frequent and Fowler (2002) identified three categories of respondents that can be selected to be included in a sample, but who do not complete or return questionnaires. This includes: (1) those respondents to whom the data collection procedures do not reach, thereby not providing such respondents with an opportunity to complete the survey questionnaire; (2) those respondents who requested to complete the questionnaire but refused to do so; and (3) those respondents who were not able to complete the questionnaire for several reasons including a language problem, illness, and lack of the necessary writing skills to complete a self-administered questionnaire.

Non-Response Bias

Regardless of the mode that contributed to the non-response, the likely effect of it is that it will produce a bias sample. A bias sample can be defined as a sample that is systematically different to the population from which it was drawn (Fowler, 2002). The following subsections first estimate the response rates of this research and then discuss the non-response bias in detail.

According to Fowler (2002), the nature of bias associated with a non-response depends upon the data collection method (i.e., mail, interview, or telephone). Since the data collection method of this research is postal survey, further discussion on the non-response bias is specific to postal surveys.

There are arguments that suggest that, in the instance of postal surveys, people who have particular interests in the subject matter or the research itself are more likely to return the mail questionnaire than those who are less interested (Fowler, 2002). The other consistent bias in mail surveys is that better-educated people often return completed questionnaires at a faster rate than respondents with less education. Fowler (2002) suggests that there is a lack of information to reliably predict when and how much non-response will affect the survey estimates. Therefore, efforts to ensure that the response rates reach a reasonable level, and to avoid procedures that systematically produce differences between respondents and non-respondents, are important ways to build confidence in the accuracy of survey estimates (Fowler,

2002). The following subsection explains the measures that were undertaken to reduce non-response in this research.

Reducing Non-Response

Fowler (2002) suggested the following three measures to reduce a non-response in mail surveys: (1) the layout should be clear, so it is easy to see how to proceed; (2) the questions should be attractively spaced, easy to read, and uncluttered; and (3) the response tasks should be easy to undertake. There should not be open-ended questions. The response tasks should be a check, a box or circling of a number. The three measures highlighted were followed whilst developing and validating the instrument through the exploratory survey, content validity, pre-test and pilot test. In order to evaluate the three criteria, questionnaires of the pilot study (Chapter 4) has included four explicit questions. The majority of respondents from the pilot study were satisfied with the length, layout, and easiness to read, which meant that there were minimal chances of non-response due to the nature of the data collection tool (i.e., postal survey) employed in this research.

Correcting Non-Response

Since non-response is inevitable, Fowler (2002) has suggested three approaches including proxy respondents and resurveying a sample of non-respondents in order to minimise the resulting error contributed due to the non-response. First, an approach that collected data from proxy respondents was considered inappropriate for this research. This is because such an approach is suitable for the interview and telephone data collection methods where any other member of a household can replace designated respondents. Therefore, the second approach, which involves resurveying a sample of non-respondents, was considered appropriate for this research.

In the resurveying approach, a sample of non-respondents should be re-contacted, either employing the same data collection method (i.e., mail) or, if the research budget allows, replacing the postal approach with a telephone survey (Fowler, 2002). Due to the following three reasons, it was decided to contact the non-respondents utilising the same data collection method, i.e., mail. The first and foremost reason for re-contacting the non-respondents via mail was the length of the survey instrument. The length of the survey instrument was six pages, which consisted of 41 Likert scale type questions, 41 usage related questions, 20 impact related questions, and a number of demographic questions. Such a questionnaire was not appropriate for data collection via the telephone method. This is because respondents prefer not to answer a long questionnaire over the telephone (Fowler, 2002). The second reason was that the sample frame utilised to select the respondents only consisted of the postal addresses of the respondents and not their telephone numbers or email addresses. The third and final reason was that this research is a student project constrained by resources that did not allow replacing a mail survey with a telephone survey in order to re-contact the sample of non-respondents. More details of resurveying the non-respondents are provided in the next section -response rate estimation- and the *t*-test results to examine if the responses of non-respondents differ from those of the respondents are provided in Chapter 5.

Response Rates

The response rate is a way of determining the success of a data collection effort and also obtaining an initial idea about the quality of the collected data. Fowler (2004) defined the response rate as the number of obtained responses divided by the number of sampled respondents, including all respondents in the study population who were sent the survey, but who did not respond (Fowler, 2002). The response rate of this study is calculated following this definition.

Of the overall 1600 questionnaires sent, 383 replies were received within the specified period. Of these, 358 questionnaires were usable and 25 were both undeliverable and incomplete questionnaires. This implies that a response rate of 22.4 percent was obtained. To test the response bias, 200 questionnaires were sent to randomly selected non-respondents from the original sample in mid-March 2005. Of this, 40 questionnaire replies were received that included 38 usable and two partially completed questionnaires. The findings obtained from the response bias test are presented in Chapter 5 (Table 5.1), which illustrates that there were no significant differences in the number of variables between the original respondents and a sample of non-respondents.

Justification for Choosing Postal Mail as a Data Collection Method

The recommended data collection methods for a survey research approach include postal mail, telephone, personal interview, group administration and the Internet (Cornford & Smithson, 1996; Fowler, 2002; Straub et al., 2005). According to Fowler (2002), the selection of a data collection method is a matter of complex decisions as it is based on a number of issues such as sampling, question form, question content, response rate, costs, available facilities and length of data collection. In turn, these factors are unique to the context of a particular study (Fowler, 2002). For example, if the collected data requires asking open questions, then it is more appropriate to employ face-to-face interviews than other methods. However, if questions are closed ended in nature and available resources are limited, then a postal questionnaire would be a better choice than other methods.

In terms of sampling, it has been suggested that the choice of data collection method should be based on reliability and the comprehensiveness of information that a sample frame offers. If the mailing addresses in a sample frame are not complete and updated, then it is not appropriate to employ postal mail as a data collection method. Such guidelines also apply to other data collection methods such as telephone (Fowler, 2002). Since the sample frame of 'UK-Info Disk V11' provided this research with a comprehensive and reliable mailing list that is updated annually, it was considered appropriate to employ mail as a data collection method. With regards to the population characteristics, it is suggested that if the reading and writing skills of the target population are low then other methods of data collection, such as the telephone or face-to-face interview are more appropriate than a self-administered questionnaire (Fowler, 2002). Since the adult literacy rate in the UK is 99 percent (economywatch. com), the reading and writing skills were not considered to be a barrier for selecting a self-administered questionnaire as a data collection method.

The self-administered method of data collection (i.e., mail) was also preferred over the interview and telephone due to ease of contact. This is because many people are busy in their daily lives and work schedules and due to this, the researcher may have encountered problems when arranging a suitable time for face-to-face or telephone interviews (Fowler, 2002). In comparison, if the contact information is correct, questionnaires can reach respondents who can then respond at a time that is convenient for them. In terms of the question format, Fowler (2002) suggests that self-administered procedures can have an advantage if the instruments comprise only closed ended questions that can be answered by simply ticking a box. When a researcher wants to ask many questions that are similar in form, then having an interviewer face-to-face or over the phone reading a long list can be awkward and tedious (Fowler, 2002). This was most relevant to this research as there were many similar types of closed ended questions that required answers. It was therefore considered more appropriate to employ the self-administered mail method rather than the telephone or face-to-face interview.

The nature of the questionnaire's contents can also influence the answers obtained from respondents. Respondents do not want to provide answers to sensitive questions over the telephone or in an interview. For this reason also, the self-administered method was found to be the most suitable data collection method (Fowler, 2002) for this research. In addition to this, other important factors that affect the data collection method are the costs and available facilities for data collection. Self-administered mail is considered to be less expensive than telephone or face-to-face interviews and requires minor involvement of additional people and other resources (Fowler, 2002). Since this research was a student project, the available resources and faculties were highly limited. Therefore, a self-administered questionnaire via mail was considered to be the most appropriate method to collect large amounts of data with the available resources and facilities.

After careful consideration of all these factors, and in order to collect random data from the target population, a self-administered questionnaire via mail was considered to be the most appropriate data collection method. In summation, the reasoning for using the self administered questionnaire was that: it addresses the issue of reliability of information by reducing and eliminating differences in the way by which the questions are asked (Cornford & Smithson, 1996); it requires relatively low costs of administration; it could be accomplished with minimal facilities; it provides access to widely dispersed samples; respondents have time to provide thoughtful answers; it helps in asking questions with long or complex response categories; it allows the asking of similar repeated questions, and also the respondents do not have to share answers with interviewers (Fowler, 2002). Therefore, the final questionnaires were sent using the postal service as a larger sample population was obtained in a cost-effective manner (Fowler, 2002). A cover letter and a self-addressed prepaid return envelope were also administered to a total of 1600 heads of households in the UK. This activity was undertaken in the period between January and March 2005. In the middle of March 2005, questionnaires were also sent to 200 non-respondents in order to assess the non-response bias.

Instrument Development and Validation

According to Fowler (2002), 'a defining property of social survey is that answers to questions are used as a measure which is a critical dimension of the quality of survey estimates.' This critical dimension depends upon reliability (i.e., providing consistent measures in comparable situations) and validity (i.e., answers correspond to what they intend to measure) of questions asked to survey respondents. Therefore, both the issue of reliability and validity of the research instrument are of utmost importance for this survey-based research (Fowler, 2002; Straub et al., 2004). Due to the criticality of the instrument in the precision of survey estimates, Straub et al., (2004) recommended that if a previously validated instrument is available for efficiency reasons, researchers should prefer utilising it rather than developing a new one. However, at the same time, researchers should not avoid previous validation controversies and, if significant changes are made in the existing instrument, it is most important to revalidate the content, construct and reliability of the modified instrument (Straub et al., 2004).

In a situation where no existing instrument is available, and where the development of a new instrument for established theoretical constructs and testing of the robustness is required, then all validities must be applied in greater detail (Straub et al., 2004).

According to Straub et al., (2004), this step is the 'heart of the demonstration of the usefulness of the new instrument' and represents 'a major contribution to scientific practice in the field' (Straub et al., 2004). Although the constructs utilised in this research belong to established theories and models, they require the development of new measures for this research. This is because the unit of analysis for this research —broadband—is different to previous technologies such as the PC, for which there is an existing instrument. Therefore, examining broadband in the household context demands creating new items or making significant changes in the existing items and employing subsequent validating measures.

Given the emphasis of the impact of a survey instrument on the reliability and validity of survey estimates or findings, it was decided to develop and validate the survey instrument for this research before proceeding to data collection. Since developing and validating an instrument is a long and stepwise process that includes exploratory survey, content validity, pre-test, pilot test, and confirmatory test, the next two chapters (Chapters 4 and 5) are devoted to providing a description of its development and validation. To avoid any repetition, no further details on this issue are provided within this section.

Data Analysis

The collated data was analysed using SPSS version 11.5. The reason for selecting the SPSS statistical package is that it facilitates the calculation of all essential statistics, such as descriptive statistics, reliability test, factor analysis, t-test, discriminant analysis, ANOVA and linear and logistic regression analysis required for data analysis and presenting findings. Furthermore, SPSS is easily available and user friendly so it can be learnt within a short period of time. An added reason for using this particular statistical package is that a number of books are available to familiarise oneself with the SPSS application to present and interpret the data.

Statistical Techniques for Validity Testing

Straub et al., (2004) recommended that a new survey instrument should be validated employing statistical techniques such as a reliability test—in order to confirm the internal consistency of measures—and factor analysis—in order to confirm the construct validity, including both convergent and discriminant validity (Straub et al., 2004). According to the recommended guidelines, a survey instrument possesses a high internal consistency (i.e., it is reliable) if the estimated Cronbach's alpha is above 0.70. Construct validity (both discriminant and convergent) exists if the latent root criterion (i.e., eigenvalue) is equal to or above 1, with a loading of at least 0.40; and no cross loading of items above 0.40. (Straub et al., 2004). Following these guidelines, the specified statistical techniques were employed to validate the survey instrument of this research (Chapters 4 and 5).

Statistical Techniques for Testing Relationships

In order to explain the relationship between the independent and dependent variables to test the conceptual model of broadband adoption, linear and logistic regression analysis was utilised. The purpose of performing linear regression analysis is to examine whether significant relationships exist between the independent variables (i.e., attitudinal, normative and control constructs) and dependent variable (behavioural intention). Multiple linear regression is a commonly-used technique to explain the relationship if the nature of both the independent and dependent variables is ordinal or scale (Brace et al., 2003; Davis et al., 1989; Oh et al., 2003; Stevens, 1996; Taylor & Todd, 1995). However, linear regression cannot be applied if the dependent variable is nominal or categorical in nature (Brace et al., 2003). The suggested alternative for this situation is logistic regression analysis, which allows for testing the relationship even if the dependent variable is nominal in nature (Brace et al., 2003). Logistic regression analysis is utilised in this research to explain the relationship between the aggregate measure of independent variables (i.e., behavioural intention and facilitating conditions resources) and the categorical dependent variable (i.e., broadband adoption).

Statistical Techniques for Testing Differences

In order to analyse nominal variables, such as demographics, usage rate, and impact of broadband, the calculation of the response frequencies and percentages was undertaken. The reason for using the aforesaid statistics is due to previous information systems (IS) researchers employing such tools for analysis and to present the research findings using response frequencies and percentages (Venkatesh & Brown, 2001; Webster, 1998). To test the statistical significance of nominal variables such as demographic differences of the adopters and non-adopters of broadband, the chi-square (χ^2) test was considered to be the most appropriate method (Brace et al., 2003).

Overall Scale Construction and Parametric Test for Difference

If all items for a construct are internally consistent (i.e., illustrate high reliability) and load on one factor in the factor analysis (i.e., demonstrate construct validity), then they can be utilised to construct a scale (i.e., aggregate measure) in the following two ways (Moore & Benbasat, 1991). The first is to construct a scale that involves summing or averaging the mean of the items that load highly on a factor (Gorsuch, 1988; Moore & Benbasat, 1991). The second is to construct a scale (i.e., aggregate measure) that necessitates considering the score of factors (Moore & Benbasat, 1991). Moore and Benbasat (1991) argued that since the relative weight of an item in a scale is based on its loading on the factor, its scores may be considered more exact than averaging means.

However, employing factor scores for constructing scales (i.e., aggregate measures) is the less preferred method (Moore & Benbasat, 1991). This is because factor scores are often less interpretable and generalisable than using the first approach that involves summing or averaging the mean of items (Moore & Benbasat, 1991). Since a number of studies (Brown et al., 2002; Karahanna et al., 1999; Koufaris, 2002; Moore & Benbasat, 1991; Oh et al., 2003; Olson & Boyer, 2003; Taylor & Todd, 1995) have employed averaging the mean of items as a means of constructing aggregate measures, and its application is reported to be entirely adequate (Moore & Benbasat, 1991; Tabachnik & Fidell, 1989), averaging responses to the individual items will be utilised to develop aggregate measures for each of the constructs in this research. Once the scale is created, it will be in a ratio instead of being ordinal, and then it will be appropriate to apply a parametric test (t-test and ANOVA) to examine the differences. Such an approach was followed by Karahanna et al. (1999) who applied a parametric test (t-test and ANOVA) on the constructed aggregate measures by averaging the mean of individual items.

It is appropriate to apply an independent t-test to this research in order to determine whether two means can be obtained from two independent respondent groups that are significantly different from each other (Brace et al., 2003; Hinton et al., 2004). In this research, groups may constitute a broadband and narrowband respondent group in order to examine the mean differences with regards to a connection type, or a male group and a female group when examining the mean differences with regards to gender. A t-test will also be utilised to test the response bias and the effect of ordering the questionnaire items.

When more than two conditions or groups of an independent variable are compared, ANOVA is more appropriate to apply than a t-test (Brace et al., 2003; Hinton et al., 2004). It is relevant to apply ANOVA to determine whether means that are obtained from more than two independent respondent groups are significantly different from each other (Brace et al., 2003; Hinton et al., 2004). In this research, ANOVA will be applied to test the scale mean differences when test variables possess more than two independent groups.

Summary

This chapter provided an overview of the research approaches that have been utilised within the IS field and then selected an appropriate research approach for guiding this particular research. To validate and understand the conceptual framework, it was found that a quantitative research would be more appropriate than a qualitative one.

An overview of the underlying epistemology was provided in order to decide that positivism should be the philosophical foundation for this research. Following this, an overview discussion on the various issues on the available research approaches in the IS field and a justification for the selection of the survey as a research approach was provided. Once it was decided that a survey is an appropriate approach to conduct this research, a detailed account of the various aspects of the survey approach was offered. It was found that for the purpose of this research, it was appropriate to employ the UK-Info Disk V11 as a sampling frame, and stratified random sampling as a basis for sample selection.

The data collection tool used in this research was the postal method (i.e., mail). The reasons for this selection were also provided in a detailed manner. Issues relating to data analysis were then discussed in detail. It was concluded that a number of statistical techniques such as factor analysis, t-test, ANOVA, χ^2 test, discriminant analysis, linear and logistics regression analysis are appropriate to be utilised for data analysis purposes. This chapter thoroughly covered two of the three essential components of survey research approach. However, the third—instrument development—was briefly introduced. The following chapters (Chapters 4 and 5) will describe the development and validation of the survey instrument that is considered essential for a reliable data collection.

References

Brace, N., Kemp, R., & Snelgar, R. (2003). *SPSS for psychologists: A guide to data analysis using SPSS for windows.* New York: Palgrave Macmillan.

Brown, S. A., Massey, A. P., Montoya-Weiss, M. M., & Burkman, J. R. (2002). Do I really have to? User acceptance of mandated technology. *European Journal of Information Systems, 11*(4), 267-282.

Cornford, T., & Smithson, S. (1996). *Project research in information systems: A student's guide.* London: Macmillan Press Ltd.

Davis, F. D. (1989). Perceived usefulness, perceived ease of use, and user acceptance of information technology. *MIS Quarterly, 13*, 319-340.

Farhoomand, A. F. (1992). Scientific progress of management information systems. In R.D. Galliers (Ed.), *Information systems research: Issues, methods and practical guidelines.* Oxford: Blackwell Scientific.

Fowler, F. J., Jr. (2002). *Survey research methods.* London: Sage Publications.

Galliers, R. D., & Land, F. F. (1987). Choosing an appropriate information systems research methodology. *Communications of the ACM, 30*(11), 900-902.

Galliers, R. D. (Ed.). (1992). Choosing Information systems research approaches. In *Information systems research: Issues, methods and practical guidelines.* Oxford: Blackwell Scientific.

Gilbert, N. (2001). *Researching social life.* London: Sage Publications.

Gorsuch, R. L. (1988). Exploratory factor analysis. In J. R. Nesselroade & R. B. Cattell (Eds.), *Handbook of multivariate experimental psychology.* New York: Plenum Press.

Hinton, P. R., Brownlow, C., McMurray, I., & Cozens, B. (2004). *SPSS explained.* East Sussex, UK: Routledge.

Kaplan, B., & Maxwell, J. A. (1994). Qualitative research methods for evaluating computer information systems. In J. G. Anderson, C. E. Aydin, & S. J. Jay (Eds.), *Evaluating health care information systems: Methods and applications.* Thousand Oaks, CA: Sage.

Karahanna, E., Straub, D. W., & Chervany, N. L. (1999). Information technology adoption across time: A cross-sectional comparison of pre-adoption and post-adoption beliefs. *MIS Quarterly, 23*(2), 183-213.

Koufaris, M. (2002). Applying the technology acceptance model and flow theory to online consumer behavior. *Information Systems Research, 13*(2), 205-223.

Mingers, J. (2001). Combining IS research methods: Towards a pluralist methodology. *Information Systems Research, 12*(3), 240-259.

Mingers, J. (2003). The paucity of multi-method research: A review of the information systems literature. *Information Systems Journal, 13*(3), 233-249.

Moore, G. C., & Benbasat, I. (1991). Development of an instrument to measure the perceptions of adopting an information technology innovation. *Information Systems Research, 2*(3), 192-222.

Myers, M. D. (1997). Qualitative research in information systems. *MIS Quarterly, 21*(2), 241-242.

Nandhakumar, J., & Jones, M. (1997). Too close for comfort? Distance and engagement in interpretive information systems research. *Information Systems Journal, 7*, 109-131.

Oh, S., Ahn, J., & Kim, B. (2003). Adoption of broadband Internet in Korea: The role of experience in building attitude. *Journal of Information Technology, 18*(4), 267-280.

Olson, J. R., & Boyer, K. K. (2003). Factors influencing the utilization of Internet purchasing in small organizations. *Journal of Operations Management, 21*, 225-245.

Orlikowski, W. J., & Baroudi, J. J. (1991). Studying information technology in organizations: Research approaches and assumptions. *Information Systems Research, 2*(1), 1-28.

Rice, C. (1997). *Understanding customers.* Oxford: Butter worth-Heinemann.

Straub, D. W., Boudreau, M-C., & Gefen, D. (2004). Validation guidelines for IS positivist research. *Communications of the Association for Information Systems, 13,* 380-427.

Straub, D. W., Gefen, D., & Boudreau, M. C. (2005). Quantitative research. In D. Avison & J. Pries-Heje (Ed.), *Research in information systems: A handbook for research supervisors and their students.* Amsterdam: Elsevier.

Stevens, J. (1996). *Applied multivariate statistics for the social sciences.* NJ: Lawrence Erlbaum Associates.

Tabachnick, B. G., & Fidell, L. S. (1989). *Using multivariate statistics.* New York: Harper and Row.

Taylor, S., & Todd, P. A. (1995). Understanding information technology usage: A test of competing models. *Information Systems Research, 6*(1), 44-176.

Venkatesh, V., & Brown, S. (2001). A longitudinal investigation of personal computers in homes: Adoption determinants and emerging challenges. *MIS Quarterly, 25*(1), 71-102.

Venkatesh, V., Morris, M. G., Davis, B. G., & Davis, F. D. (2003). User acceptance of information technology: Toward a unified view. *MIS Quarterly, 27*(3), 425-478.

Walsham, G. (1993). *Interpreting information systems in organizations.* Chichester: Wiley.

Walsham, G. (1995a). The emergence of interpretivism in IS research. *Information Systems Research, 6*(4), 376-394.

Walsham, G. (1995b). Interpretive case studies in IS research: Nature and method. *European Journal of Information System, 4*(2), 74-81.

Webster, J. (1998). Desktop videoconferencing: experiences of complete users, wary users, and non-users. *MIS Quarterly, 22*(3), 257-286.

Appendix: IS Research Approaches

Minger (2003)	Galliers (1992)	Straub et al., (2005)	
	Positivist		
		QPR	Non-QPR
Observation (passive), measurements, and (statistical) analysis	Laboratory experiment	Lab Experiment	Math Modeling
Experiments	Field experiment	Field Experiment	Group feed-back
Survey, questionnaire, or instrument	Survey	Field Study	Participative Research
Case study	Case study	Adaptive Experiment	Case study
	Theorem proof	Opinion research	Philosophical research
	Forecasting	Archival research	
Simulation	Simulation	Free & Experimental Simulation	
	Interpretivist		
Interviews	Subjective/argumentative		
Qualitative content analysis	Reviews		
Ethnography	Action research		
Grounded theory	Descriptive/interpretive		
Participant observation	Futures research		
	Role/game playing		
	Methods involving interventions (Critical Research)		
Action research			
Critical theory			
Consultancy			

(Adapted from Galliers, 1992; Mingers, 2003; Straub et al., 2005)

Chapter 4

Development of Survey Instrument:
Exploratory Survey and Content Validity

Abstract

Chapter 2 described the proposed conceptual model that is used to understand the adoption, usage, and impact of broadband from the consumer perspective. Chapter 3 described the appropriate research approach for testing the hypotheses and to validate the proposed conceptual model. From Chapter 3 it was concluded that the survey research approach is an appropriate method to investigate the issue of broadband diffusion. Further suggestions that were provided in Chapter 3 are before conducting the final data collection a reliable survey instrument should be developed and validated. Validating an instrument is a critical step before testing a conceptual model (Boudreau et al., 2001; Straub et al., 2004). This is due to the rigour of the findings and interpretations in positivist research that are based on the solid validation of the instruments used to gather data (Boudreau et al., 2001; Straub et al., 2004). Therefore, this chapter aims to describe the development of a survey instrument designed to investigate broadband adoption, usage, and impact within UK households. By undertaking the following three stages, this led to the development of a reliable instrument: (1) to explain broadband adoption behaviour some initial factors were identified from the literature and then a decision upon how to determine them in an exploratory survey approach needed to be made; (2) content validation was performed on the itemed pools that resulted from the exploratory survey. The purpose of this step was to confirm the representativeness of items to a particular construct domain and, finally, (3) a pre-test and a pilot test were conducted utilising the obtained instrument after content validation was undertaken in order to confirm the reliability of the measures. The next section briefly re-introduces the conceptual model and provides a list of the constructs included in the various stages of the validation process. Following this, an overview of the instrument development process is provided. Then the first stage of the validation process (i.e., the exploratory survey) is presented and discussed. This is followed by the content validation process. The instrument testing process that includes the pre-test and pilot-test is described before presenting the summary of the chapter.

Conceptual Model

Although the conceptual model has been described in Chapter 2, a brief account of the constructs is provided in this section. The constructs included in this study were adapted from the model of the adoption of technology in households (MATH) (utilitarian outcomes, hedonic outcomes, and knowledge) (Venkatesh & Browns, 2001), diffusion of innovations (relative advantage) (Rogers, 1995), and the theory of planned behaviour (TPB) (behavioural intention, social influence, facilitating conditions resources, self-efficacy) (Ajzen, 1991; Taylor & Todd, 1995).

The proposed model assumed that the dependent variable, behavioural intention towards broadband adoption is influenced by several independent variables that include the attitudinal (relative advantage, utilitarian outcomes, and hedonic outcomes), normative (primary influence), control factors (knowledge, self-efficacy and facilitating conditions resources), and demographic variables (age, gender, income, education, and occupation). Although a detailed discussion of each construct is not possible within the scope of this chapter, a list of the constructs included at each stage of validation is illustrated in Table 4.1. These constructs were defined (see Table 2.2) and discussed in Chapter 2.

Straub et al. (2004) suggested that if content is adapted from an existing instrument then there is less need to validate it. However, if there are changes made in an instrument then the adapted measures should be subjected to a rigorous validation process (Straub et al., 2004).

Following the reasoning provided by Straub et al. (2004), the adoption-related items are to be subjected to all four stages of validation. The usage items were not included in the first stage, but subjected to validation in the remaining three stages. In the case of the usage of broadband, there was no modification required in content but it was not certain if the list is exhaustive or not; therefore, this construct was considered from the content validity stage and onwards. Since the impact items are adapted from a previous instrument (Horrigan & Rainie, 2002) that examined the impact of broadband on consumers from the USA, they were only included in the last stage of validation (i.e., pilot test) (Straub et al., 2004).

Instrument Development Process

Since this is the first attempt at developing a research instrument for the purpose of investigating broadband adoption, usage, and impacts amongst household consumers, a number of stages were followed that included the selection and creation of items, an exploratory survey, content validity, a pre-test, and a pilot test. Straub et al. (2004) argued that there is a lack of a standard validation approach. Various studies

Table 4.1. List of constructs included in the various stages of instrument validation

Constructs	Exploratory Survey	Content Validity	Pre-Test	Pilot Test
Behavioural Intention	No	Yes	Yes	Yes
BISP	No	Yes	Yes	Yes
Relative Advantage	Yes	Yes	Yes	Yes
Utilitarian Outcomes	Yes	Yes	Yes	Yes
Hedonic Outcomes	Yes	Yes	Yes	Yes
Service Quality	No	Yes	Yes	Yes
Primary Influence	Yes	Yes	Yes	Yes
Secondary Influence	Yes	Yes	Yes	Yes
Facilitating Conditions Resources	Yes	Yes	Yes	Yes
Knowledge	Yes	Yes	Yes	Yes
Self-efficacy	Yes	Yes	Yes	Yes
Usage (Online Activities)	No	Yes	Yes	Yes
Impact	No	No	No	Yes

have employed different methods and techniques of validation (Straub et al., 2004). Therefore, keeping that in mind, this study adapted a validation approach similar to the IS studies (for example, Davis, 1989; Moore & Benbasat, 1991) that have focused upon the instrument development process and have been widely cited by researchers. Since these studies developed an instrument in a stepwise manner, this research also followed a similar approach of developing and validating the instrument in several stages, including an exploratory survey, content validity, pre-test, and pilot test. A detailed discussion on each stage is now provided.

Stage 1: The Exploratory Survey

Research Method

The purpose of the exploratory survey was to determine the items or factors affecting the adoption and non-adoption behaviours. Following this, in subsequent stages, the important factors that could be modified to measure perceptions were identified. This was undertaken bearing in mind that the broadband adopters were asked to answer only the attitudinal and normative-related questions. In contrast, the broadband non-adopters were asked to complete questions related to only the control constructs. A number of appropriate items were collated, created, and modified to suit the definition of the related attitudinal, normative, and control constructs. However, it is important to mention here that one attitudinal construct—service quality—and its related items were not included at this stage.

Further explanations on this issue are provided at the end of this section. A definition for all the constructs is provided in Table 2.2. A list of constructs included in the exploratory survey and related items are provided in Table 4.1 and Appendix 4.1 respectively.

The data for the exploratory survey were collected from the household consumers living in the local vicinity of the London borough of Hillingdon. The selection of the target population was made according to the availability of the sample frame. Since a reliable sample frame—that is the electoral register—was not easily available for the whole of London or the UK population, it was decided to conduct a survey within the London borough of Hillingdon. The structure of the sample frame (the electoral register) necessitated the adoption of a stratified random sampling approach that collected the representative data from the target population. The whole locality was divided into various wards and sub-wards in the electoral register. The sample size for each sub-ward was determined according to the total population. Following this, unique, random numbers for each sub-ward were generated using the research randomizer software. The respondents' corresponding random numbers were then selected for data collection from the sample frame (the electoral register).

In order to collect random data for the target population and within a limited time frame and resources, a self-administered questionnaire was considered to be an appropriate primary survey instrument. This is because it addresses the issue of reliability of information by reducing and eliminating the differences the way that the questions are asked (Cornford & Smithson, 1996).

The questionnaire used in the exploratory survey contained a total of 13 questions. These questions were divided into three broad categories: (1) multiple choice questions addressing the social attributes (demographic variables) including age, gender, education, and income; (2) Likert scale (1-5) questions that were designed to address the issues related to the factors of broadband adoption, and (3) an open-ended question that asked respondents if they would like to mention any other factor that was not included in the questionnaire regarding their decision of subscribing to broadband or not.

The final questionnaire was sent using the postal service. A cover letter and a pre-paid return envelope were administered to a total of 700 households in the London borough of Hillingdon during August and September, 2003. The collected data was analysed using SPSS version 11.5. The analysis was focused upon calculating the importance of the attitudinal, normative and control factors utilising the means and standard deviations. In order to measure the internal consistency of the items, the reliability of scale (Cronbach's α) was also calculated.

Findings from the Exploratory Survey

Of the 700 questionnaires distributed, 200 replies were received within the specified time period. Of these, 172 questionnaires were usable for the analysis, whilst 28 were both undelivered and uncompleted questionnaires. This yielded a response rate of 25.6 percent.

Descriptive Statistics: Attitudinal Constructs

Amongst the attitudinal constructs, relative advantage was rated most strongly, followed by utilitarian outcomes (Appendix 4.1). The hedonic outcome construct was considered least important with a mean score of 2.34 and a standard deviation of 1.85. Of the four relative advantage items, faster access was rated most strongly (M=4.88, SD=0.33), followed by un-metered access (M=4.49, SD=0.83). The provision of a free home phone line was considered least important (M=3.22, SD=1.7)]. The remaining two items, which were always-on access (M=4.13, SD=1.0) and faster file downloads (M=4.12, SD=1.25) were rated as almost equally important (Appendix 4.1).

The second strongest construct was utilitarian outcome that had a rating of M=2.89, SD=0.87 amongst the attitudinal category that consisted of six items. Amongst the six items, obtaining educational material (M=3.31, SD=1.22) was considered to be the most important reason for subscribing to broadband. This was followed by performing job related tasks (M=3.05, SD=1.55) and communication with family and friends (M=3.04, SD=1.46). The factor of helping children with homework (M=2.33, SD=1.49)] was considered least important within this category. The other two items of utilitarian outcomes, performing home business (M=2.62, SD=1.45), and performing personal and household activities (M=2.96, SD=1.35) were rated as moderate (Appendix 4.1). The third construct from the attitudinal category—hedonic outcomes—was rated below average (M=2.34, SD=1.850) on a 1-5 point Likert scale. This construct consisted of only two items, of which one item was for entertainment, such as downloading, viewing, and listening to music and movies, which scored an above average score (M=2.90, SD=1.45), and the second was for playing online games and was rated below average (M=1.78, SD=1.15) (Appendix 4.1).

Descriptive Statistics: Normative Constructs

The normative dimension consisted of only two constructs: primary and secondary influences. Three items represented the first construct—primary influence—(M=2.47, SD=1.7) and the second construct, secondary influence, consisted of only one item that was rated slightly above average (M=2.60, SD=1.78). Amongst the items of the primary influence construct, influence from family members and relatives was considered most important (M=2.56, SD=1.88), followed by influence from friends (M=2.38, SD=1.77). The least rated item was influence from kids (M=2.28, SD=1.94) (Appendix 4.1).

Descriptive Statistics: Control Constructs

The control category was composed of mainly three constructs: facilitating conditions, knowledge, and skill. However, there were two items placed in the category of other items. The other items did not fit in with the definition of the above three constructs; therefore, it was decided to examine them separately. The construct facilitating conditions was represented by two items. The findings, illustrated in Appendix 4.1, suggest that high monthly cost was a key barrier preventing consumers from subscribing to broadband (Mean= 4.25 on a five point scale and SD=1.18). The second item of this construct related to the cost of purchasing a new computer or upgrading the existing one (M=2.94, SD=1.64), which was also considered to be an important factor for not adopting broadband. The second construct in the control category, knowledge, consisted of two items and was considered overall as less important (M=2.35, SD=1.37). A lack of knowledge about the usage and benefits of broadband were considered more important (M=2.33, SD=141) than the second item that referred to the lack of knowledge on broadband (M=2.28, SD=1.37) (Appendix 4.1).

A single item represented the third construct of self-efficacy, which was the lack of skills when using the computer and the Internet (M=1.95, SD=1.41). Self-efficacy was considered less influential in terms of preventing respondents from subscribing to broadband (Appendix 4.1).

The first item that was placed in the other items category was the lack of needs when subscribing to broadband. This was considered quite important in inhibiting the adoption of broadband (M=3.83, SD=1.25). The second item within the other items category was the lack of content and applications with the existing broadband packages (M=2.55, SD=1.25), which was rated less influential than the first one (Appendix 4.1).

Reliability Test

To test the internal consistency of measures, a reliability test was performed. The Cronbach's α values for all but two constructs are listed in Appendix 4.1. The value of reliability (alpha) varies for the different constructs. Since the two constructs secondary influence and self-efficacy were represented only by one item each, it was not possible to calculate their reliability. Of the remaining seven constructs, only three constructs attained an alpha above 0.60 (Appendix 4.1), which is the minimum acceptable level for the exploratory study (Straub et al., 2004, p. 411).

Amongst the attitudinal constructs, the minimum value of reliability (0.24) was associated with the relative advantage construct. This was followed by the hedonic outcomes measure (0.54). The maximum value of reliability within this category was 0.66, which was for the utilitarian outcomes construct. This suggests that only one attitudinal construct satisfied the criteria of internal consistency. For the two normative constructs, only one construct—primary influence—allowed the Cronbach's α value to be calculated. The construct attained a reliability of 0.84, which is considered acceptable (Straub et al., 2004, p. 411).

Amongst the four control constructs, knowledge attained the highest value (α=0.94) for reliability. This was followed by facilitating conditions with α=0.50. The third construct of self-efficacy did not satisfy the criteria for calculating the α value. The fourth construct achieved a α value of 0.34. As for the attitudinal and normative constructs, only one construct of the control category knowledge satisfied the criteria of internal consistency.

Limitations and Further Improvement

For Stage 1, the major limitation was the lack of previous studies that had developed and utilised the scale to measure broadband adoption for the attitudinal, normative and control constructs. Such unavailability compelled the researchers to develop an instrument from the initial stages, which instigated several issues. The following are issues that emerged from the exploratory survey and were dealt with in two stages: content validity and instrument testing.

Need for Additional Constructs

A number of respondents with narrowband connections commented that they were not satisfied with their quality of service, which included the required speed, security, and customer or technical support. If these issues were not dealt with in a narrowband context, consumers stated that they would switch to a broadband connection. However, the majority of the respondents with narrowband connections also commented that they were satisfied with the quality of service that they were receiving from their current service providers, hence they would not switch to a broadband connection. Some of the respondents with a broadband connection provided similar comments to respondents with narrowband and stated that they were also not receiving the quality of speed and support that was affirmed before subscribing to the services. As a result, they had considered transferring to other Internet service providers (ISPs). Considering these comments from the survey respondents, it was felt appropriate to include a new construct for the purpose of measuring the attitudes towards the service of quality being received from the current ISPs. Therefore, a new construct named 'service quality' was included in the content validity stage.

The other limitation to the exploratory survey was the lack of a dependent variable that could be utilised to measure the intentions of respondents when subscribing to broadband. This variable is also important when examining how independent variables (attitudinal, normative, and control constructs) affect a respondent's intention to adopt broadband in the home. This necessitated the researchers to include an additional construct called 'behavioural intention' from the theory of planned behaviour (Ajzen, 1991).

Need for New Items

Since the two constructs secondary influence and self-efficacy were composed of only one item each (Appendix 4.1), this caused problems when calculating the reliability of the construct. Therefore, this limitation was also considered during the content validity test and one more additional item was added to each construct.

Problem of Low Reliability

From the exploratory study findings and the previous discussion it was found that although the estimated mean value of many constructs (e.g., relative advantage) was high, the reliability (alpha) was low (Appendix 4.1). In contrast, the estimated mean value of the social influence construct was low, but its reliability value was higher than any other construct. These variations in estimated values necessitated a further validation of the instrument content. In order to further validate the survey instrument and to determine how representative the items for a particular construct were (Straub et al., 2004), it was decided that the content validity approach would be employed. The IS literature suggests that it is an important and highly recommended practice to conduct content validity in instances of new instrument development and also even if existing scales have to be applied for the establishment of any new object (Straub et al., 2004). Since this condition applies to this research as well, this was an added reason for conducting content validity. Content validity followed the exploratory study and its application in this research is described in the next section.

Stage 2: Content Validation

Content validity is defined as the 'degree to which items in an instrument reflect the content universe to which the instrument will be generalised' (Straub et al., 2004). Generally, content validity involves the evaluation of a new survey instrument. This is to ensure that the survey instrument that aims to measure broadband adoption, usage, and impact includes all the essential items and eliminates undesirable items within a particular construct's domain (Boudreau et al., 2001; Kitchenham & Pfleeger, 2002; Lewis, 1995; Straub et al., 2004).

Although only two approaches that comprise judgements and statistics are available when determining content validity, their application is unique to each study (Emory & Cooper, 1991; Torkzadeh & Dhillon, 2002). The application of content validity differs in terms of when it is utilised, how it is conducted, and how many experts evaluate the content. The judgement approach to establish content validity involves literature reviews and then follow-ups with evaluation by expert judges or panels. The validation of the items is based on a high degree of consensus amongst expert panels or judges on the items in question; therefore, it is judgemental in nature (Boudreau et al., 2001; Davis, 1989; Kitchenham & Pfleeger, 2002; Moore & Benbasat, 1991; Smith et al., 1996; Storey et al., 2000; Straub et al., 2004; Torkzadeh & Dhillon, 2002). Lawshe (1975) introduced an empirical or quantitative approach of content validity. This approach involves estimating the statistical validity ratio (Lawshe, 1975; Lewis et al., 1995).

The procedure of the judgemental approach of content validity requires researchers to be present with experts in order to facilitate validation. Therefore, it is also sometimes known as 'face validity' (Wacker, 2004). However, it is not always possible to have many experts of a particular research topic at one location, which was the case in this research. Alternatively, a quantitative approach allows researchers to send content validity questionnaires to experts working in different locations. Therefore, distance is not a problem faced by research. In order to perform content validity for broadband diffusion research, a quantitative approach was considered to be more suitable in comparison to a judgemental approach (Lawshe, 1975; Lewis et al., 1995). Since broadband diffusion studies are still emerging in nature, there are still few academic experts. Furthermore, the experts are located in different places. Therefore, bearing these issues in mind, the quantitative approach pursued in this research is described in the next section.

Research Method

The content validity of the broadband diffusion instruments was performed employing a quantitative approach (Lawshe, 1975). In terms of IS research, such an approach has been successfully applied when validating information resource management instruments (Lewis et al., 1995). In order to validate the content of the constructs, the quantitative approach (Lawshe, 1975; Lewis et al., 1995) was undertaken in the following manner:

- First, relevant items from the existing literature on technology adoption and diffusion were identified. This led to the construction of the questions and the content validity questionnaire.

- Second, a content evaluation panel, consisting of experts from academia and/or industry who were related to the specific research area, was selected.

- Third, each member of the panel was then provided with the questionnaire formed in the initial stage. The panel members were requested to respond independently to each item in relation to a particular construct on a three-point scale where: "1 = not necessary," "2 = useful but not essential" and "3 = essential."

- Fourth, the responses from the overall panelists were then pooled. This step also included counting responses that indicated 'essential' for each item.

- Fifth, the content validity ratio (CVR) for each item was estimated utilising the formula $CVR=(n-N/2)/(N/2)$ (Lawshe, 1975), where N is the total number of respondents and n is the frequency count of the number of panelists rating the item as "3 = essential."

- Finally, the CVR values obtained for each item were examined for their significance employing the standard table provided by Lawshe (1975). If the estimated CVR value was equal to or above the standard value, then the item was accepted, otherwise it was eliminated. The significance level (standard value) depended upon the number of experts rating the item. The minimum number of experts required to rate each item should be five. The value of CVR ranged from 0 to 1(Lawshe, 1975; Lewis et al., 1995).

These steps were followed in order to evaluate the content of the broadband diffusion survey instruments. A sample of items for each construct was identified employing an exhaustive review of literature on generic technology adoption topics, broadband adoption, and diffusion. The literature review led to the identification of 75 adoption entries and 35 entries for the usage related constructs. A content validity questionnaire was then generated that comprised the definition of the constructs and associated items on a 1-3 scale.

The experts who had been identified for content validity purposes and who engaged in broadband diffusion-related research were then approached. A total of 12 academic experts were identified on the basis of publications in peer-reviewed journals and leading conferences (ten experts) or their engagement to the research area related to broadband diffusion (two experts). The questionnaire was then sent to the experts via email so as to expedite the process. The purpose of the study and instructions to complete the questionnaire were detailed in the covering email. The experts were asked to rate each item in relation to the different constructs of broadband diffusion on a three-point scale: "1 = not necessary;" "2 = useful but not essential;" "3 = essential." They were also requested to provide comments if the items were not understandable, required re-wording or if new entries needed to be added. Responses from all the experts were then collated by counting the numbers of ratings that indicated "essential" for each item.

Following this process, the CVR was estimated and evaluated for a statistical significance level of 0.05 by employing Lawshe's (1975) method that was mentioned in the prior paragraph. This process was undertaken for each item. For entries where the 0.05 significance level was not achieved, these were eliminated. The list of constructs—along with their associated CVR values—is presented in Appendix 4.2 and 4.3.

Findings from Content Validation

The estimated values for the content validity ratio of all the items are presented in Appendix 4.2 for the adoption constructs and Appendix 4.3 for the usage constructs. A summary of the CVR that is derived from Appendix 4.2 and 4.3 is provided in Tables 4.2 and 4.3. The CVR questionnaire comprised a total of 110 items. Of the 110 items, 75 belonged to adoption and 35 to usage related constructs. The findings presented in Table 4.2 illustrate that, of the 75 items from the adoption domain, the majority of respondents considered 41 items important for inclusion, as the CVR value was significant at the 0.05 level. All 35 usage-related items were considered essential. Additionally, the respondents suggested six new online activities to include. Therefore, a total of 41 online activities were included in the next step (Table 4.7).

Table 4.3 illustrates the overall items, average CVR and average mean for each construct. The average CVR value for the 12 constructs was between the maximum value of 0.98 and minimum value of 0.57 at the 0.05 level of statistical significance. This illustrates that the constructs possess a high level of content validity, which means that the items are representative of a construct universe (Table 4.3).

The experts also provided a number of suggestions regarding the re-wording and decomposition of some of the items and the addition of a few new items. It was suggested that the BI3 (Table 4.6) items should be reworded when measuring the behavioural intention that is used to determine the extent of the use of broadband. Furthermore, the experts also commented that it would be useful to include a new item that measured the behavioural intention for determining whether a consumer continues with the current ISPs or switches to a new one. With regards to the utilitarian outcomes construct, it was suggested that item UO5 (Table 4.6) should be divided into two. This was required because information searches and online shopping are two different activities and measuring both of them with the same item could create ambiguities in the findings. Similarly, the experts suggested that item HO1 should be decomposed into two items by separating music and movies into two different items. Item HO4 was also eliminated (Table 4.6). This was because e-greetings are utilised more as a communication rather than as an entertainment tool. For the service quality items SQ11 and SQ13 (Table 4.6), the same issue was being determined, but in opposite directions; therefore, the experts advised that it would be better to drop item SQ13. The content validity experts suggested that although the two items FC1 and FC3 of the facilitating condition construct (Table 4.6) are useful to include in the questionnaire, they are not suitable for measuring perception. The preceding two items represented an actual situation and not a perception. Further, the items could be measured utilising a dichotomous variable, that is, yes/no. Considering the suggestions of the experts, these two items were removed from the facilitating condition construct. Accordingly, the suggestions were incor-

Table 4.2. Summary of content validity ratio

CVR	AI	UI
0.90-0.99	5	31
0.80-0.89	10	4
0.70-0.79	0	0
0.60-0.69	11	0
0.50-0.59	15	0
0.40-0.49*	0	0
0.30-0.39*	6	0
0.20-0.29*	0	0
0.10-0.19*	6	0
0-0.09*	13	0
Total	66	35
RLH	9	0
Grand Total	75	35

*Legend: *= Not Significant, RLH = Items that rated essential by less then half participants, AI= Adoption Items, UI= Usage Items, II= Impact Items*

Table 4.3. Summary of constructs, TI, SI, ACVR, and AM

Constructs	TI	SI	ACVR	AM
Behavioural Intention	3	2	0.83	2.83
Relative Advantage	9	4	0.63	2.79
Utilitarian Outcomes	14	9	0.74	2.84
Hedonic Outcomes	4	3	0.71	2.86
Service Quality	13	5	0.73	2.80
Primary Influences	4	3	0.56	2.78
Secondary Influences	4	2	0.75	2.88
Requisite Knowledge	6	3	0.61	2.69
Self-efficacy	7	3	0.56	2.61
Facilitating conditions	9	5	0.70	2.79
Usage	35	35	0.98	2.99

Legend: TI= Total number of items, SI= Number of significant Items, ACVR = Average content validity ratio, AM= Average mean porated along with the suggestions obtained from the pre-test of the instruments in the resulting instrument.

The experts also commented upon the direction of items that were being measured by particular constructs. More discussion on this issue and the improvements made after the suggestions are provided within the sub-section entitled 'Pre-Test.' The experts also agreed that for the final questionnaire, the 1-7 scale would be more suitable in comparison to the 1-5 scale. This is because the 1-7 scale values are widely spread in comparison to the 1-5 scale, which meant that the respondents had more choices from which to select a response. This prevents respondent bias as respondents can then select a neutral value. Therefore, 1-7 is considered to be the most suitable Likert scale for the final study.

The experts who evaluated the content of the instrument belonged to several countries: the UK, Denmark, USA, Australia, and Canada. Therefore, the content of the questionnaire is not only specific to the UK, but also to the countries from which the experts were from themselves. It was considered essential to have a mix of the countries, as a comparative study of the pre-test and validation of the questionnaire in a number of contexts—such as the USA, Australia, Canada, and EU member states—would then be possible.

The findings suggest that the content validity experts rated those items that were adopted from the exploratory survey conducted in stage 1 and from the previous study on broadband adoption in South Korea as essential (Oh et al., 2003). In contrast, the items adopted from the general technology adoption studies (Davis, 1989; Taylor & Todd, 1995) were rated but considered as non-essential. Therefore, the content validity practice confirms that the items investigated in the exploratory studies are important to understand consumers' broadband adoption behaviour.

Limitations Encountered During Content Validation

The following three limitations were encountered whilst conducting the content validity for the broadband adoption survey instruments: first, locating the experts relating to the specific research area, and second, conducting content validity with the experts located in different places, and third, the length of the content validity instruments. Since research in broadband adoption and diffusion from the demand perspective is novel, researchers involved in examining this issue are few in numbers. This is particularly true in the instance of locating experts in a country such as the UK. However, the problem was overcome by considering researchers located in other countries, such as the USA, Australia, Canada, and Denmark.

The second limitation was distance and the lack of face-to-face interaction with experts. This problem became obvious when the majority of the researchers initially perceived that the questionnaire sent to them was the final instrument. This created confusion when evaluating the questionnaire content. To overcome this problem, several e-mails were sent to each expert's queries to clarify the context of content validity.

The length of the instrument extended as the content validity questionnaire comprised a definition of each construct and related items. For example, the content validity instruments in this study were 10 pages long. Initially, the length of the questionnaire discouraged many experts from participating in the content evaluation exercise. However, after repeated contact and requests, the researchers agreed to assist with this exercise.

To complete the instrument development process that will be utilised to investigate broadband diffusion, the next step was to conduct a pre-test and pilot-test of the questionnaire using respondents from a target population.

Stage 3: Instrument Testing

Pre-Test

A pre-test of the resulting instrument was conducted with 20 respondents ranging from the broadband industry (3), an IT manager of a county council (1), academics and researchers (10) and household consumers (6). In addition to providing responses, the respondents were asked to determine whether the questions were grammatically correct, understandable, and to suggest further improvements. The pre-test participants suggested that all the items for one construct should be measured in the same direction. The content validity experts also repeated this suggestion. Therefore, this particular issue was considered carefully and such changes were made wherever required. The following is an example of such a change:

The items for the facilitating condition construct 'before pre-test' were:

1.   I cannot subscribe to broadband at home because it is too costly to purchase a new computer or to upgrade my old computer.

2.   My annual household income level is enough to afford subscribing to broadband.

3.   It is too costly for me to subscribe to broadband at its current subscription fee.

4.   I would be able to subscribe to broadband if I wanted to.

The items for the facilitating condition construct 'after pre-test' were:

1. It is not too costly to purchase a new computer or to upgrade my old computer.

2. My annual household income level is enough to afford subscribing to broadband.

3. It is not too costly for me to subscribe to broadband at its current subscription fee.

4. I would be able to subscribe to broadband if I wanted to.

The suggestions that were provided by the content validity experts regarding the re-wording and modifying of items were all implemented at this stage and this is reflected in the pilot questionnaire. Another issue that the respondents commented upon at the pre-test stage was the length of the questionnaire. Prior to the questionnaire being sent to the pre-test participants, the questionnaire was 11 pages long. The respondents expressed concern about the length of the questionnaire. They suggested that in its current form, the questionnaire length was extensive and could lead to a low response rate. Their suggestion was that the length of the questionnaire should be reduced but without losing the measurement content. Bearing in mind the comments of the participants, the total length of the questionnaire was reduced to six pages. This was achieved by making two changes that consisted of the structure and format of the questions. For example:

1. In the pre-test questionnaire the options for the Likert scale questions were arranged in two rows. However, after the comments for each item, the Likert scale options were managed within one row. This assisted in reducing the length of the questionnaire, e.g.:

The arrangement of the Likert Scale items in the pre-test questionnaire:
Broadband has an advantage over dial-up/narrowband because it offers faster access to Internet:

☐ 7= Extremely agree   ☐ 5= Slightly agree   ☐ 3= Slightly disagree

☐ 6= Quite agree ☐ 4= Neutral ☐ 2= Quite disagree ☐ 1= Extremely disagree

Arrangement of the Likert Scale items in the final questionnaire:
Broadband has an advantage over dial-up/narrowband because it offers faster access to the Internet:

Extremely disagree   □ 1 □ 2 □ 3 □ 4 □ 5 □ 6 □ 7   **Extremely agree**

2.   The second change was made to the usage part of the questionnaire. Both the content validity experts and pre-test participants suggested this change. In the pre-test questionnaire, usage of each online activity was a separate question that required a lot of space, which meant a longer questionnaire. However, following the suggestions, the final questionnaire consisted of these online activities being arranged in a tabular form. An example is provided below:

Online services	From Home		From Work Place, Internet Café, Library		Would you like to use it in the future?	
E-mail	Yes	No	Yes	No	Yes	No
Instant messaging	Yes	No	Yes	No	Yes	No

The arrangement of the online activities in the pre-test questionnaire:

How often do you access the following online services from home?

1.   E-mail

□ Several times a day   □ 3-5 days a week   □ Once every few weeks

□ About once a day   □ 1-2 days a week   □ Less often   □ Never

2.   Instant messaging

□ Several times a day   □ 3-5 days a week   □ Once every few weeks

□ About once a day   □ 1-2 days a week   □ Less often   □ Never

The arrangement of the online activities in the final questionnaire:

Besides the aforementioned changes and a few spelling and typographical errors, the respondents from the pre-test studies supported the content of the questionnaire. After incorporating all the suggested changes by the content validity experts and pre-test participants, the length of the resultant questionnaire was reduced to six pages. More details regarding the contents of the final questionnaire are provided in the following sub-section.

Pilot-Test

Research Method

The final stage of the instrument development process was a pilot test of the questionnaire using respondents whose backgrounds were similar to the final study's target population. The respondents were selected utilising the 'UK-Info Disk V11' database. The database comprised the nationwide addresses of households in the UK. The primary aim of the test was to ensure that the various scales demonstrated the appropriate levels of reliability. The pilot test also indicated to the researchers an estimate of the actual response rates. Furthermore, the pilot test identified difficulties that respondents could face when completing the questionnaire. For example, some of the difficulties included determining whether the diction of the questionnaire and the accompanying instructions were comprehensive enough for completion of the questionnaire (Moore & Benbasat, 1991).

From the pre-test, a survey instrument resulted that was six pages long and consisted of an overall total of 17 questions. The questions were divided into four categories: (1) multiple type questions examining the demographics of the respondents (questions 1-6), Internet connection types, frequency and duration of Internet access on a daily basis (question 8-12); (2) yes/no questions that determined the location of the Internet at home or elsewhere (question 7), accessibility to various (total of 41) online activities (question 16); (3) Likert scale questions to assess the perception of the adopters and non-adopters of broadband (question 13); and (4) to assess the impact of broadband upon individuals' time allocation patterns, respondents were questioned about their use of the Internet. That is, whether usage of broadband had increased, decreased or had no impact, upon the amount of time spent on the 20 various daily life activities (question 17).

Four other questions were also asked in order to determine respondents understanding of the questionnaire. The four questions were: (1) Is the length of the questionnaire appropriate? (2) Are the questions understandable? (3) Is the layout of the questionnaire acceptable? (4) How long did it take the respondent to complete the questionnaire? The responses of the respondents to these four questions are summarised in Table 4.5.

The pilot questionnaires were sent to an overall total of 200 respondents via the postal service in the month of December, 2004. A covering letter that included the definitions of broadband, narrowband, metered and un-metered, and a self-addressed prepaid return envelope were also included with the questionnaire. A total of 42 replies were received within the specified time limit. Of the replies, there were 40 usable responses and the remaining two replies were incomplete; therefore they offered no relevance and were excluded from the analysis. Hence, a response rate of 20 percent was obtained.

Table 4.4. Summary of statistics obtained from pilot-test (N = 40)

Constructs	Number of Items	Scale Mean	Scale SD	RELIABILITY Cronbach's Alpha (α)	Type (Hinton et al., 2004, pp 364)
Behavioural Intention	2	5.5	1.7	0.95	**Excellent Reliability**
*BISP	1	3.54	2.15	---	---
Relative Advantage	4	6.3	0.7	0.75	High Reliability
Utilitarian Outcomes	10	5.7	1.0	0.91	**Excellent Reliability**
Hedonic Outcomes	4	3.7	1.7	0.88	High Reliability
Service Quality	4	4.2	1.3	0.78	High Reliability
Primary Influence	3	4.5	1.6	0.88	High Reliability
Secondary Influence	2	3.6	1.9	0.95	**Excellent Reliability**
Facilitating Conditions Resources	4	5.0	1.2	0.72	High Reliability
Knowledge	3	5.7	1.3	0.85	High Reliability
Self-efficacy	3	6.2	1.2	0.94	**Excellent Reliability**
**Usage (Online Activities)	41	---	---	---	---
**Impact	20	---	---	---	---

LEGEND: BISP=Behavioural Intention to change subscriber, **SD**= Standard Deviations

*Reliability is not estimated since construct is composed of only one item

** Reliability is not estimated since variables are nominal (categorical) in nature

Findings from the Pilot Test

Of the overall 40 responses, 21 (52.5 percent) respondents had narrowband (dial up) connections. Further, seven (17.5 percent) of them had metered narrowband, whilst 14 (35 percent) had un-metered narrowband. There were 13 (32.5 percent) respondents who had broadband. The broadband replies also had categories and the resulting replies are as follows. Eight (20 percent) respondents had DSL, four (10 percent) had a cable modem and one (2.5 percent) had a wireless connection. The remaining six (15 percent) respondents did not access the Internet from home using either narrowband or broadband. Considering the diversity of the connection types, this sample of 40 respondents was considered to be a good sample for the purposes of pilot testing the instrument, since it consisted of various types of adopters and non-adopters of narrowband and broadband.

The mean, standard deviation (SD), and reliability (alpha) obtained from the pilot-test are presented in Table 4.4. The descriptive statistics (mean and standard

Table 4.5. Respondent perception of survey instrument   (N = 40)

Questions	Frequency		Percent	
	Yes	No	Yes	No
1. Is the length of the questionnaire appropriate?	35	5	87.5	12.5
2. Are the questions understandable?	39	1	97.5	2.5
3. Is the layout of the questionnaire OK?	37	3	92.5	7.5
	Time require to complete questionnaire			
4. How long did it take to complete the questionnaire?	10 Min.	15 Min.	20 Min.	25 Min
Frequency	15	12	12	1
Percent	37.5	30.0	30.0	2.5

deviation) illustrated in Table 4.4 suggest that amongst the attitudinal constructs, the relative advantage construct is highly rated, with the lowest standard deviations at a seven point Likert scale and the hedonic outcome being poorly rated. Amongst the normative constructs, primary influence was considered to be more important than secondary influence. The descriptive statistics for the control constructs suggest that the skill construct is highly rated. This means that the majority of respondents including the adopters and non-adopters possess skills required for broadband adoption and use. Although a brief summary of the descriptive statistics is provided here, an in-depth analysis of the significance of the factors is not provided, as this does not fit with the scope of this chapter. The following sub-sections discuss the test of instrument reliability.

The statistics obtained from the reliability analysis confirm the internal consistency of the measure. Cronbach's α for this test varies between 0.95 for the three constructs of behavioural intention, secondary influence and self-efficacy, and 0.72 for facilitating conditions resources. All the constructs included in the survey possess Cronbach's alpha above 0.70. The IS literature offers advice that in order to satisfy the internal consistency criteria, Cronbach's α should be above 0.60 for an exploratory survey, and 0.70 for a confirmatory study (Straub et al., 2004, pp 411). Therefore, following this 'rule of thumb,' the obtained values are acceptable. This confirms that the measurement is internally consistent and possesses an appropriate reliability level. Hinton et al. (2004) have suggested four cut-off points for reliability, which includes excellent reliability (0.90 and above), high reliability (0.70-0.90), moderate reliability (0.50-0.70), and low reliability (0.50 and below) (Hinton et al., 2004, pp 364). According to those of the 10 constructs, four possess excellent reliability and the remaining six constructs demonstrate high reliability. None of the constructs demonstrated a moderate or low reliability (Table 4.4).

Table 4.6. List of constructs and items to examine broadband adoption

---

1. BEHAVIOURAL INTENTION (BI) TO ADOPT BROADBAND

BI1: I intend to subscribe to (or continue my current subscription) broadband in the future

BI3: I intend to use (or intend to continue use) broadband Internet service in the future

2. BEHAVIOURAL INTENTION TO CHANGE SERVICE PROVIDER (BISP)

BI2: I intend to continue my current subscription but will change the current service provider

3. RELATIVE ADVANTAGE (RA)

RA1: Broadband has an advantage over dial-up because it offers faster access to Internet

RA2: Broadband has an advantage over dial-up because it provides faster download of files from Internet

RA3: Broadband has an advantage over dial-up because it offers an always-on access to Internet

RA4: Broadband has an advantage over dial-up because it frees up the phone line whilst connected to the Internet

4. UTILITARIAN OUTCOMES (UO)

UO1: Broadband can be useful to find educational materials and accessing library resources at home

UO2: Broadband can be useful for distance learning

UO3: Broadband can be helpful to perform work/job-related tasks at home

UO4: Broadband will help me communicate better via email, chat, Web cam

UO5: Broadband can help in performing personal and household activities i.e. *online shopping*

UO6: Broadband can help in performing personal and household activities i.e. *information search*

UO7: Broadband can be helpful to establish and operate a home business

UO8: Broadband can help children to do their homework

UO9: Subscribing to broadband is compatible with most aspects of my everyday life

UO10: Overall broadband will be useful to me and other members in the family

5. HEDONIC OUTCOMES (HO)

HO1: I will enjoy using broadband to listen to and download music

HO2: I will enjoy using broadband to watch to and download movies

HO3: I will enjoy using broadband to play online games

HO4: I will enjoy using broadband to play online gambling/casino

6. SERVICE QUALITY

SQ1: I am satisfied with the speed of Internet access obtained from my current service providers

SQ2: I am satisfied with the security measures provided with Internet access obtained from my current service providers

SQ3: I obtained satisfactory customer/technical support from my current service providers

SQ4: The overall service quality of my current Internet connection is satisfactory

---

continued on following page

Table 4.6. continued

---

7. PRIMARY INFLUENCE

PI1: My friends think that I should subscribe to (or continue the current subscription) broadband at home

PI2: My colleagues think that I should subscribe to (or continue the current subscription) broadband

PI3: My family members think that I should subscribe to (or continue the current subscription) to broadband

8. SECONDARY INFLUENCE

SI1: TV and radio advertising encourages me to try broadband

SI2: Newspaper advertising encourages me to try broadband

9. FACILITATING CONDITIONS RESOURCES

FCR1: My annual household income level is enough to afford subscribing to broadband

FCR2: It is not too costly to purchase a new computer or to upgrade my old computer

FCR3: It is not too costly for me to subscribe to broadband at its current subscription fee

FCR4: I would be able to subscribe to broadband if I wanted to

10. KNOWLEDGE

K1: I do not have difficulty in explaining why adopting broadband may be beneficial

K2: I know how broadband is different from dial-up/narrowband Internet

K3: I know the benefits that broadband offer and cannot be obtained by dial-up/narrowband

11. SELF-EFFICACY

SE1: I would feel comfortable using the Internet on my own

SE2: Learning to operate the Internet is easy for me

SE3: I clearly understand how to use the Internet

Comparing the alpha values obtained from the exploratory survey (Appendix 4.1) and pilot-test (Table 4.4), it is clearly suggested that after conducting the content validation and pre-test, the scale reliability has improved for all the constructs. This improvement provides evidence that content validation is not only required to test the representativeness of the items for a construct domain, but is also helpful in improving the internal consistency of measures.

There were no comments obtained from the pilot respondents for improving the questionnaire. Only a few respondents suggested that it may be useful to provide definitions of terms, such as broadband, narrowband, metered and un-metered. Table 4.5 illustrate that of the 40 pilot respondents, 35 (87.5 percent) agreed that the length of the questionnaire was appropriate. Of the 40, 39 (97.5 percent) respondents found the questions to be understandable and 37 (92.5 percent) respondents indicated that the layout of the questionnaire was appropriate. To determine the time taken

Table 4.7. List of online activities for examining broadband usage

SN	Online services	SN	Online services
1	Email	22	Listen to the radio station
2	Instant messaging	23	Watch movies (downloading/streaming)
3	Online Chat	24	Undertake online banking
4	Online News	25	Online bill paying
5	Job related research	26	Purchase a product
6	Look for product info	27	Purchase a travel service
7	Research for school or training	28	Online auctions e.g. e-bay
8	Look for travel information	29	Purchase groceries (household goods)
9	Look for medical information	30	Buy/sell stocks (online share trading)
10	Share computer files	31	Play lottery
11	Create content (e.g. Web pages)	32	Obtain information on hobby
12	Store/display/develop photos	33	Use it for fun e.g. Web surfing
13	Store files on the Internet	34	Play online game
14	Download games	35	View or visit Adult content Websites
15	Download video	36	Video conferencing
16	Download pictures	37	Voice over Internet (VoIP)
17	Download music	38	Online dating and matrimonial services
18	Download movie	39	Online lectures
19	Download free software	40	Collaboration with schoolmates
20	Video streaming/downloading	41	Accessing e-government services
21	Listen to music (streaming/MP3)		

to complete the questionnaire, the participants were offered four options consisting of 10, 15, 20, and 25 minutes. The findings presented in Table 4.5 illustrate that 67.5 percent of the respondents took between 10 (37.5 percent respondents) to 15 minutes (30.0 percent respondents) to complete the questionnaire. As the outcome of the pilot findings in mind were positive, it was decided that the administered questionnaire for the pilot study did not require any further changes and was considered appropriate for the final survey.

Final Survey Instrument

The list of items for the adoption constructs utilised for the final data collection is presented in Table 4.6. However, the complete survey instrument (i.e., the question-

Table 4.8. List of daily life activities for examining broadband impact

SN	Activities	SN	
1	Watching television/cable/satellite	11	Time spent on hobbies
2	Shopping in stores	12	Time spent on sleeping
3	Working at home	13	Time spent alone (doing nothing)
4	Reading newspapers/books/magazines	14	Studying
5	Working in the office	15	Household work
6	Commuting in traffic	16	Receiving/ making phone calls
7	Spending time with family	17	Doing charity and social works
8	Spending time with friends	18	Outdoor recreation (DIY, pet care)
9	Attending social events	19	Outdoor entertainment (concerts, cinema)
10	Time spent on sport	20	Visiting or meeting friends or relatives

naire) that resulted from the pilot-test is provided in Appendix 4.4. A total of 40 items formed the survey content of the final instrument and belonged to 11 different constructs that included both the independent and dependent variables. Of the remaining 11, the two constructs behavioural intention (BI) and secondary social influence (SI) were composed of two items each. Only the construct behavioural intention to change the service provider (BISP) was represented by only one item. The three constructs that are primary influence (PI), knowledge (K), and self-efficacy (SE) consisted of three items each. Four constructs, namely relative advantage (RA), hedonic outcomes (HO), service quality (SQ), and facilitating conditions resources (FCR) were composed of four items each. The utilitarian outcomes construct was (UO) represented by 10 items (Table 4.6).

The usage part of the questionnaire was composed of 41 online activities, which are listed in Table 4.7. The categorical variables that included duration and frequency of Internet access are not presented in Table 4.7 but were included in the questionnaire (Appendix 4.9). To examine if Internet use by broadband and narrowband consumers had an impact upon their daily life, 20 different activities were included within the final questionnaire (Appendix 4.4). These activities are listed in Table 4.8. Apart from the adoption, usage, and impact related variables listed above, five demographic variables (age, gender, education, income, and occupation) were also included in the questionnaire (Appendix 4.4). The theoretical justification and relevance for including these variables are provided in Chapter 3.

Summary

This chapter described the development process for a survey instrument that was utilised to examine broadband adoption within the household context. The development process was achieved in three stages: the exploratory survey, content validity, and instrument testing. The processes of each stage are summarised in the following paragraphs.

The exploratory stage included surveying the known existing instruments, choosing appropriate items, creating required new items and then determining if the selected items were appropriate enough to measure the perceptions of the adopters and non-adopters of broadband. This stage also examined the reliability (internal consistency) of the initial scale. At this stage, it was found that the majority of items were either selected from the existing instrument or were newly created items that were important enough to explain the behaviour of the adopters and non-adopters. However, the reliability of the scale was low in most cases. The exploratory stage also led the researcher to identify the new construct that was referred to as 'service quality' and some new items as well. The output of this stage of research was utilised as an input for the content validity stage.

The content validity stage included the creation of new items for each construct and then the validation of their representativeness utilising a quantitative approach. These new items were created utilising the items obtained from the exploratory stage and also by re-surveying the literature and selecting the relevant items. In order to achieve the representativeness of items, several experts working on broadband-related issues evaluated the newly created items. This led the researchers to calculate a content validity ratio that was the basis of the exclusion or inclusion of the items. The outcome of this stage was the inclusion of the representative items and the exclusion of non-related items. The reliability of all the 10 scales improved after conducting the content validity process. This demonstrated the importance of performing content validation for the increased reliability of the scale and also the representativeness of the items.

The instrument testing stage was sub-divided into two stages: the pre-test and pilot test. The purpose of the pre-test was to obtain feedback on the instrument from the respondents, and to improve the wording of items. The purpose of the pilot test was to confirm the reliability of items. The findings obtained from the pilot-test demonstrated an acceptable level of reliability for all the constructs. The final output of the three-stage instrument development process that culminated from the pilot-test is a parsimonious 40-item instrument, consisting of 11 scales, all with a high level of reliability. The final instrument will be utilised to investigate the behavioural intentions of household consumers when adopting broadband and also its usage and impact.

The next chapter (Chapter 5) will analyse and present the findings obtained from the final data collection that was conducted on a nationwide basis. Chapter 5 first estimates the response rate and then conducts the response bias test. This is then followed by a re-examination of the reliability of the instrument and then construct validity is performed by conducting principal component analysis (PCA). A test was also conducted to assess the biases that may have been introduced due to the ordering of questions.

References

Ajzen, I. (1991). The theory of planned behaviour. *Organisational Behaviour and Human Decision Processes, 50,* 179-211.

Cornford, T., & Smithson, S. (1996). *Project research in information systems: A student's guide.* London: Macmillan Press Ltd.

Davis, F. D. (1989). Perceived usefulness, perceived ease of use, and user acceptance of information technology. *MIS Quarterly, 13,* 319-340.

Boudreau, M., Gefen, D., & Straub, D. (2001). Validation in IS research: A state-of-the art assessment. *MIS Quarterly, 25*(1), 1-23.

Emory, C. W., & Cooper, D. R. (1991). *Business research methods.* Boston: Irwin.

Hinton, P. R., Brownlow, C., McMurray, I., & Cozens, B. (2004). *SPSS explained.* East Sussex, UK: Routledge.

Horrigan, J. B., & Rainie, L. (2002). *The broadband difference: How online Americans' behaviour changes with high-speed Internet connections at home.* Retrieved September 20, 2003, from http://www.pewinternet.org/pdfs/PIP_Broadband_Report.pdf

Kitchenham, B. A., & Pfleeger, S. L. (2002). Principles of survey research part 2: Designing a survey. *Software Engineering Notes, 27*(1), 18-20.

Lawshe, C. H. (1975). A quantitative approach to content validity. *Personnel Psychology, 28,* 563-575.

Lewis, B. R., Snyder, C. A., & Rainer, K. R. Jr. (1995). An empirical assessment of the information resources management construct. *Journal of Management Information Systems, 12*(1), 199-223.

Moore, G. C., & Benbasat, I. (1991). Development of an instrument to measure the perceptions of adopting an information technology innovation. *Information Systems Research, 2*(3), 192-222.

Oh, S., Ahn, J., & Kim, B. (2003). Adoption of broadband internet in Korea: The role of experience in building attitude. *Journal of Information Technology*, *18*(4), 267-280.

Rogers, E. M. (1995). *Diffusion of innovations.* New York: Free Press.

Smith, H. J., Milberg, S. J., & Burke, S. J. (1996). Information privacy: Measuring individuals' concern about organizational practices. *MIS Quarterly, 20*(2), 167-197.

Storey, V. C., Straub, D. W., Stewart, K. A., & Welke, R. J. (2000). A conceptual investigation of the e-commerce industry. *Communications of the ACM, 43*(7), 117-123.

Straub, D. W., Boudreau, M-C, & Gefen, D. (2004). Validation guidelines for IS positivist research. *Communications of the Association for Information Systems, 13*, 380-427.

Taylor, S., & Todd, P. A. (1995). Understanding information technology usage: A test of competing models. *Information Systems Research, 6*(1), 44-176.

Torkzadeh, G., & Dhillon, G. (2002). Measuring factors that influence the success of Internet commerce. *Information Systems Research, 13*(2), 187-204.

Venkatesh, V., & Brown, S. (2001). A longitudinal investigation of personal computers in homes: Adoption determinants and emerging challenges. *MIS Quarterly*, *25*(1), 71-102.

Wacker, J. W. (2004). A theory of formal conceptual definitions: developing theory-building measurement instruments. *Journal of Operations Management, 22*(6), 629-650.

Appendix A

Appendix 4.9. Summary of statistics obtained from exploratory survey

Adoption/Rejection factors		Mean	SD	R (α)
INDICATORS FOR BROADBAND ADOPTION (TOTAL RESPONSES N=51)				
ATTITUDINAL FACTORS				
1. RELATIVE ADVANTAGE	5	**4.17**	**0.56**	**0.235**
Faster access to the Internet		4.88	0.33	
Always-on access to the Internet		4.13	1.0	
Free home phone line		3.22	1.7	
Un-metered access to Internet		4.49	0.83	
To download files faster		4.12	1.25	
2. UTILITARIAN OUTCOMES	6	**2.89**	**.87**	**0.66**
To perform job-related tasks		3.06	1.55	
To find educational/research materials		3.31	1.22	
To perform home business		2.63	1.45	
To help with children's homework		2.33	1.49	
To perform the personal & household		2.96	1.35	
To communicate with family, friends and relatives		3.03	1.46	
3. HEDONIC OUTCOMES	2	**2.34**	**1.85**	**0.54**
To play online games		1.78	1.15	
For entertainment such as music and movies		2.90	1.45	
NORMATIVE FACTORS				
1. PRIMARY INFLUENCE	3	**2.47**	**1.7**	**0.84**
Influence from family members and relatives		2.56	1.88	
Influence from friends		2.38	1.77	
Influence from Kids		2.28	1.94	
2. SECONDARY INFLUENCE	1	---	---	---
Influence from TV/News advert		2.60	1.78	
INDICATORS FOR BROADBAND NON-ADOPTION (N=78)				
CONTROL FACTORS				
1. FACILITATING CONDITIONS RESOURCES	2	**3.60**	**1.17**	**0.50**
Cost of purchasing/upgrading the computer		2.94	1.64	
High monthly cost of broadband subscription		4.25	1.18	
2. KNOWLEDGE	2	**2.35**	**1.37**	**0.94**
Lack of knowledge about broadband		2.28	1.43	
Lack of knowledge about broadband usage and benefits		2.33	1.41	
3. SELF-EFFICACY	1	---	---	---
Lack of skills to use computer and Internet		1.95	1.41	
4. OTHER ITEMS	2	**3.19**	**0.97**	**0.34**
Lack of need to subscribe the broadband		3.83	1.25	
Lack of content/applications with broadband		2.55	1.25	

[Legend: TI= Total number of items, SD= Standard deviation, R (α)= Reliability (Cronbach's α)]

Appendix 4.10. Estimation of content validity ratio (CVR) for adoption items (N=12)

SN	I. No.	3	2	1	n	Mean	CVR	SN	I. No.	3	2	1	n	Mean	CVR
1	BI1	12	0	0	12	3	1	39	SQ8	6	2	4	6	2.17	0
2	BI2	5	4	3	5	2.167	-0.17	40	SQ9	12	0	0	12	3	1
3	BI3	10	0	2	10	2.67	0.67	41	SQ10	4	2	6	4	1.83	-0.33
4	RA1	10	2	0	10	2.83	0.67	42	SQ11	9	1	2	9	2.58	0.5
5	RA2	9	3	0	9	2.75	0.5	43	SQ12	6	1	6	6	2.17	0
6	RA3	6	4	2	6	2.33	0	44	SQ13	9	1	2	9	2.58	0.5
7	RA4	11	1	0	11	2.92	0.83	45	PI1	9	3	0	9	2.75	0.5
8	RA5	9	2	1	9	2.67	0.5	46	P12	10	2	0	10	2.83	0.67
9	RA6	6	4	2	6	2.33	0	47	PI3	9	3	0	9	2.75	0.5
10	RA7	8	2	2	8	2.5	0.33	48	PI4	7	3	2	7	2.42	0.17
11	RA8	6	3	3	6	2.25	0	49	SI1	11	1	0	11	2.92	0.83
12	RA9	8	2	2	8	2.5	0.33	50	SI2	6	3	3	6	2.25	0
13	UO1	12	0	0	12	3	1	51	SI3	10	2	0	10	2.83	0.67
14	UO2	11	1	0	11	2.92	0.83	52	SI4	6	2	4	6	2.17	0
15	UO3	12	0	0	12	3	1	53	RK1	9	2	1	9	2.67	0.5
16	UO4	11	1	0	11	2.92	0.83	54	RK2	8	3	1	8	2.58	0.33
17	UO5	11	1	0	11	2.92	0.83	55	RK3	8	1	3	8	2.42	0.33
18	UO6	9	2	1	9	2.67	0.5	56	RK4	7	2	3	7	2.33	0.167
19	UO7	10	1	1	10	2.75	0.67	57	RK5	11	0	1	11	2.83	0.83
20	UO8	7	1	4	7	2.25	0.17	58	RK6	9	1	2	9	2.58	0.5
21	UO9	6	3	3	6	2.25	0	59	SE1	10	0	2	10	2.67	0.67
22	UO10	6	3	3	6	2.25	0	60	SE2	7	2	3	7	2.33	0.167
23	UO11	4	5	3	4	2.08	-0.33	61	SE3	9	1	2	9	2.58	0.5
24	UO12	6	2	4	6	2.17	0	62	SE4	9	1	2	9	2.58	0.5
25	UO13	5	2	5	6	2.0	-0.17	63	SE5	5	5	2	5	2.25	-0.167
26	UO14	9	2	1	9	2.67	0.5	64	SE6	7	2	3	7	2.33	0.167
27	UO15	9	2	1	9	2.67	0.5	65	SE7	7	0	5	7	2.17	0.167
28	HO1	11	1	0	11	2.92	0.83	66	FC1	11	0	1	11	2.83	0.83
29	HO2	10	2	0	10	2.83	0.67	67	FC2	10	1	1	10	2.75	0.67
30	HO3	9	3	0	9	2.75	0.5	68	FC3	11	1	0	11	2.92	0.83
31	HO4	11	1	0	11	2.92	0.83	69	FC4	8	0	4	8	2.33	0.33
32	SQ1	12	0	0	12	3	1	70	FC5	9	3	0	9	2.75	0.5
33	SQ2	6	1	5	6	2.08	0	71	FC6	6	0	6	6	2	0
34	SQ3	10	2	0	10	2.83	0.67	72	FC7	6	1	5	6	2.08	0
35	SQ4	4	3	5	4	1.92	-0.33	73	FC8	10	0	2	10	2.67	0.67
36	SQ5	5	3	4	5	2.08	-0.17	74	FC9	4	1	7	4	1.75	-0.33
37	SQ6	4	3	5	4	1.92	-0.33	75	FC10	10	2	0	10	2.83	0.67
38	SQ7	8	2	2	8	2.5	0.33								

Legend: N= Total number of experts completed the content validity questionnaire, n= number of experts rated items as essential, I. No.= Item number, CVR (content validity ratio)= n-N/2/N/2

Appendix 4.11. Estimation of content validity ratio (CVR) for usage items (N=12)

Online services	N	3	2	1	n	Mean	CVR
Email	12	12	0	0	12	3	1
Instant messaging	12	12	0	0	12	3	1
Online Chat	12	12	0	0	12	3	1
Online News	12	12	0	0	12	3	1
Job related research	12	12	0	0	12	3	1
Look for product info	12	12	0	0	12	3	1
Research for school or training	12	12	0	0	12	3	1
Look for travel information	12	12	0	0	12	3	1
Look for medical information	12	12	0	0	12	3	1
Share computer files	12	12	0	0	12	3	1
Create content (e.g. Web pages)	12	12	0	0	12	3	1
Store/display/develop photos	12	12	0	0	12	3	1
Store files on the Internet	12	12	0	0	12	3	1
Download games	12	12	0	0	12	3	1
Download video	12	12	0	0	12	3	1
Download pictures	12	12	0	0	12	3	1
Download music	12	12	0	0	12	3	1
Download movie	12	12	0	0	12	3	1
Download free software	12	12	0	0	12	3	1
Video streaming/downloading	12	12	0	0	12	3	1
Listen to music (streaming/MP3)	12	12	0	0	12	3	1
Listen to the radio station	12	12	0	0	12	3	1
Watch movies (downloading/streaming)	12	12	0	0	12	3	1
Undertake online banking	12	12	0	0	12	3	1
Online bill paying	12	11	0	1	11	2.83	0.83
Purchase a product	12	12	0	0	12	3	1
Purchase a travel service	12	12	0	0	12	3	1
Online auctions e.g. e-bay	12	12	0	0	12	3	1
Purchase groceries (household goods)	12	11	1	0	11	2.92	0.83
Buy/sell stocks (online share trading)	12	12	0	0	12	3	1
Play lottery	12	12	0	0	12	3	1
Obtain information on hobby	12	12	0	0	12	3	1
Use it for fun e.g. Web surfing	12	11	1	0	11	2.92	0.83
Play online game	12	11	1	0	11	2.92	0.83
View or visit Adult content Websites	12	12	0	0	12	3	1

Legend: N= Total number of experts completed the content validity questionnaire, n= number of experts rated items as essential, I. No.= Item number, CVR (content validity ratio)= n-N/2/N/2

Dear Sir/Madam,

You are kindly requested to participate in a nationwide survey research being conducted by Mr. Yogesh Dwivedi, a PhD candidate, under the supervision of Dr Jyoti Choudrie, Director of Operations, Brunel Broadband Research Centre in the School of Information Systems, Computing and Mathematics, Brunel University.

The aim of this research is to "investigate broadband adoption, usage, and impact in the UK household". The questionnaire consists of a number of questions that should take approximately 15 minutes to complete. Please tick all appropriate answers. If your answer is not displayed, then please state your answer in the "other" option category. Participation is voluntary. You may omit any questions that you do not wish to answer.

None of the information provided by the participants will be disclosed or used in any monetary, political or institutional way. Your name will not be revealed in any of the documents unless you grant permission. A code number will be used to protect your identity.

Data will be kept with the investigator and supervisor and will be destroyed after completion of this research.

If you have any questions about this study, please contact the investigators on the following address: Mr. Yogesh Dwivedi, PhD Student, School of Information Systems, Computing and Mathematics, Brunel University, Uxbridge, Middlesex UB8 3PH, United Kingdom, email: cspgykd@brunel.ac.uk, phone: (01895) 265969.

We would like to take this opportunity to thank you for your time and patience in completing this questionnaire!

Note: In this questionnaire the term dial-up/narrowband refers to the Internet connection that offers a speed below 128 Kilobits per second. Broadband refers to a high speed, always-on and un-metered Internet connection. The offered speed is above 128 Kilobits per second. The term un-metered refers to a fixed subscription fee and metered means cost per usage.

Broadband Diffusion Survey

1. Who is (with reference to household/family heads) completing the questionnaire?

- ○ Head of household
- ○ Son/daughter
- ○ Parents
- ○ Boarder/Lodger

- ○ Relative
- ○ Son/daughter-in-law
- ○ Spouse
- ○ Others (Please specify)...

2.   What age group do you belong to?

- Under 16 Years   ○ 45-54 Years
- 17-24 Years   ○ 55-64 Years
- 25-34 Years   ○ 65-74 Years
- 35-44 Years   ○ Above 75 Years

3.   Gender

- Male   ○ Female

4.   Highest level of education

- GCSE   ○ A Level   ○ Postgraduate Taught (MA, MSc)
- GNQV/Diploma   ○ Degree   ○ Postgraduate Research (PhD)

5.   What is your occupation?

- Directors, doctors, lawyers, professors
- Managers, teachers, computer programmers
- Foremen, shop assistants, office workers
- Electricians, mechanics, plumbers and other crafts)
- Machine operators, assembly, cleaning
- Pensioners, casual workers, unemployed, students
- Others (Please specify)…

6.   What is your household's annual income? (K= £1000)

- <10 K   ○ 40-49 K
- 10-19 K   ○ 50-59 K
- 20-29 K   ○ 60-69 K
- 30-39 K   ○ =>70 K

7.   Do you have Internet access at home?

- Yes (Please complete all the questions below)
- No (Please go to question 10)

8.   If you do have Internet access, what would you describe the type of Internet is that you do have?

○ Dial-up metered   ○ Broadband with DSL/ADSL

○ Dial-up un-metered ○ Broadband with CABLE MODEM

○ Wireless   ○ Other

9.   How long have you been accessing the Internet for?

○ <12 Months

○ 12-24 Months

○ 25-36 Months

○ >36 Months Others

10.   Where else do you obtain access to the Internet? (Please tick all applicable options)

○ Work place   ○ University or college

○ Public access points ○ Local library

○ Internet cafe   ○ Other (Please specify)…

11.   How often do you access the Internet (From work place, Internet café or library)?

○ Several times a day ○ 1-2 days a week

○ About once a day   ○ Once every few weeks

○ 3-5 days a week   ○ Less often

                       ○ Other

12.   How long do you spend on the Internet on a daily basis?

○ <1/2 hour   ○ >1-2 hour   ○ >3-4 hour   ○ >5-6 hour   ○ >7-8 hour

○ >9-10 hour ○ 1/2-1 hour ○ >2-3 hour   ○ >4-5 hour   ○ >6-7 hour

○ >8-9 hour   ○ Other

13. Please rate each of the following statements provided on a 1-7 point scale where:

1 = Extremely disagree   2 = Quite disagree

3 = Slightly disagree   4 = Neutral

5 = Slightly agree   6 = Quite agree

7 = Extremely agree

The following statements only represent your perception so it is alright to rate them even if you do not have broadband Internet connection at home.

BI1. I intend to subscribe to (or continue my current subscription) broadband in the future

Extremely disagree   1   2   3   4   5   6   7   **Extremely agree**

BI2. I intend to continue my current subscription but will change the current service provider

Extremely disagree   1   2   3   4   5   6   7   **Extremely agree**

BI3. I intend to use (or intend to continue use) broadband Internet service in the future

Extremely disagree   1   2   3   4   5   6   7   **Extremely agree**

RA1. Broadband has an advantage over dial-up/narrowband because it offers faster access to Internet

Extremely disagree   1   2   3   4   5   6   7   **Extremely agree**

RA2. Broadband has an advantage over dial-up because it provides faster download of files from Internet

Extremely disagree   1   2   3   4   5   6   7   **Extremely agree**

RA3. Broadband has an advantage over dial-up because it offers an always-on access to Internet

Extremely disagree   1   2   3   4   5   6   7   **Extremely agree**

RA4. Broadband has an advantage over dial-up because it frees up the phone line whilst connected to the Internet

Extremely disagree   1   2   3   4   5   6   7   **Extremely agree**

UO1. Broadband can be useful to find educational materials and accessing library resources at home

Extremely disagree   1   2   3   4   5   6   7   **Extremely agree**

UO2. Broadband can be useful for distance learning

Extremely disagree   1   2   3   4   5   6   7   **Extremely agree**

UO3. Broadband can be helpful to perform work/job-related tasks at home

Extremely disagree   1   2   3   4   5   6   7   **Extremely agree**

UO4. Broadband will help me communicate better via e-mail, chat, Web cam

Extremely disagree   1   2   3   4   5   6   7   **Extremely agree**

UO5. Broadband can help in performing personal and household activities, i.e., *online shopping*

Extremely disagree   1   2   3   4   5   6   7   **Extremely agree**

UO6. Broadband can help in performing personal and household activities, i.e., *information search*

Extremely disagree   1   2   3   4   5   6   7   **Extremely agree**

UO7. Broadband can be helpful to establish and operate a home business

Extremely disagree   1   2   3   4   5   6   7   **Extremely agree**

UO8. Broadband can help children to do their homework

Extremely disagree   1   2   3   4   5   6   7   **Extremely agree**

U9. Subscribing to broadband is compatible with most aspects of my everyday life

Extremely disagree   1   2   3   4   5   6   7   **Extremely agree**

U10. Overall broadband will be useful to me and other members in the family
Extremely disagree   1   2   3   4   5   6   7   **Extremely agree**

HO1. I will enjoy using broadband to listen to and download music
Extremely disagree   1   2   3   4   5   6   7   **Extremely agree**

HO2. I will enjoy using broadband to watch to and download movies
Extremely disagree   1   2   3   4   5   6   7   **Extremely agree**

HO3. I will enjoy using broadband to play online games
Extremely disagree   1   2   3   4   5   6   7   **Extremely agree**

HO4. I will enjoy using broadband to play online gambling/casino
Extremely disagree   1   2   3   4   5   6   7   **Extremely agree**

SQ1. I am satisfied with the speed of Internet access (dial-up or broadband) obtained from my current service providers
Extremely disagree   1   2   3   4   5   6   7   **Extremely agree**

I am satisfied with the security measures provided with Internet access (dial-up or broadband) obtained from my current service providers
Extremely disagree   1   2   3   4   5   6   7   **Extremely agree**

SQ3. I obtained satisfactory customer/technical support from my current service providers
Extremely disagree   1   2   3   4   5   6   7   **Extremely agree**

SQ4. The overall service quality of my current Internet connection is satisfactory
Extremely disagree   1   2   3   4   5   6   7   **Extremely agree**

PI1. My friends think that I should subscribe to (or continue the current subscription) broadband at home
Extremely disagree   1   2   3   4   5   6   7   **Extremely agree**

P12. My colleagues think that I should subscribe to (or continue the current subscription) broadband

Extremely disagree   1   2   3   4   5   6   7   **Extremely agree**

P13. My family members (i.e., spouse, kids) think that I should subscribe to (or continue the current subscription) to broadband

Extremely disagree   1   2   3   4   5   6   7   **Extremely agree**

SI1. TV and radio advertising encourages me to try broadband

Extremely disagree   1   2   3   4   5   6   7   **Extremely agree**

SI2. Newspaper advertising encourages me to try broadband

Extremely disagree   1   2   3   4   5   6   7   **Extremely agree**

K1. I do not have difficulty in explaining why adopting broadband may be beneficial

Extremely disagree   1   2   3   4   5   6   7   **Extremely agree**

K2. I know how broadband is different from dial-up/narrowband Internet

Extremely disagree   1   2   3   4   5   6   7   **Extremely agree**

K3. I know the benefits that broadband offer and cannot be obtained by dial-up/narrowband

Extremely disagree   1   2   3   4   5   6   7   **Extremely agree**

SK1. I would feel comfortable using the Internet on my own

Extremely disagree   1   2   3   4   5   6   7   **Extremely agree**

SK2. Learning to operate the Internet is easy for me

Extremely disagree   1   2   3   4   5   6   7   **Extremely agree**

SK3. I clearly understand how to use the Internet

Extremely disagree   1   2   3   4   5   6   7   **Extremely agree**

FCT1. My current PC is good enough to access the Internet

Extremely disagree   1   2   3   4   5   6   7   **Extremely agree**

FCT2. There is no problem of broadband Internet availability in my locality

Extremely disagree   1   2   3   4   5   6   7   **Extremely agree**

FCR1. My annual household income level is enough to afford subscribing to broadband

Extremely disagree   1   2   3   4   5   6   7   **Extremely agree**

FCR2. It is not too costly to purchase a new computer or to upgrade my old computer

Extremely disagree   1   2   3   4   5   6   7   **Extremely agree**

FCR3. It is not too costly for me to subscribe to broadband at its current subscription fee

Extremely disagree   1   2   3   4   5   6   7   **Extremely agree**

FCR4. I would be able to subscribe to broadband if I wanted to

Extremely disagree   1   2   3   4   5   6   7   **Extremely agree**

14. Do you access the following online services from home or from any other location such as, the Work Place, Internet Café, Library or Public Access Point? Please make 'Yes' or 'No' bold and underline it in both the sections.

    Also would you like to access the following services utilising home Internet connection in the future as well? Please make 'Yes' or 'No' bold and underline it in the last column.

15. Has the use of the Internet increased, decreased or had no change on the amount of time you spend on the following activities? Please highlight your answer by making it bold and underlining it.

16. Are you aware about online Government Gateway?

  ○ Yes   ○ No

17. Have you registered with online Government Gateway?

  ○ Yes   ○ No

18.   If you would like to enter the prize draw then please state your address below. If the candidate is successful, then the prize will be sent to the stated address by post.

Home Address..

E-mail ....

Thank you very much for your valuable time and patience for completing this questionnaire!

Appendix 4.13

Online services	From Home		From Work Place, Internet Café, Library		Would you like to use it in the future?	
E-mail	Yes	No	Yes	No	Yes	No
Instant messaging	Yes	No	Yes	No	Yes	No
Online Chat	Yes	No	Yes	No	Yes	No
Online News	Yes	No	Yes	No	Yes	No
Job related research	Yes	No	Yes	No	Yes	No
Look for product info	Yes	No	Yes	No	Yes	No
Research for school or training	Yes	No	Yes	No	Yes	No
Look for travel information	Yes	No	Yes	No	Yes	No
Look for medical information	Yes	No	Yes	No	Yes	No
Share computer files	Yes	No	Yes	No	Yes	No
Create content (e.g., Web pages)	Yes	No	Yes	No	Yes	No
Store/display/develop photos	Yes	No	Yes	No	Yes	No
Store files on the Internet	Yes	No	Yes	No	Yes	No
Download games	Yes	No	Yes	No	Yes	No
Download video	Yes	No	Yes	No	Yes	No
Download pictures	Yes	No	Yes	No	Yes	No
Download music	Yes	No	Yes	No	Yes	No

continued on following page

Appendix 4.13. continued

Download movie	Yes No	Yes No	Yes No
Download free software	Yes No	Yes No	Yes No
Video streaming/downloading	Yes No	Yes No	Yes No
Listen to music (streaming/MP3)	Yes No	Yes No	Yes No
Listen to the radio station	Yes No	Yes No	Yes No
Watch movies (downloading/streaming)	Yes No	Yes No	Yes No
Undertake online banking	Yes No	Yes No	Yes No
Online bill paying	Yes No	Yes No	Yes No
Purchase a product	Yes No	Yes No	Yes No
Purchase a travel service	Yes No	Yes No	Yes No
Online auctions e.g. e-bay	Yes No	Yes No	Yes No
Purchase groceries (household goods)	Yes No	Yes No	Yes No
Buy/sell stocks (online share trading)	Yes No	Yes No	Yes No
Play lottery	Yes No	Yes No	Yes No
Obtain information on hobby	Yes No	Yes No	Yes No
Use it for fun, e.g., Web surfing	Yes No	Yes No	Yes No
Play online game	Yes No	Yes No	Yes No
View or visit Adult content Web sites	Yes No	Yes No	Yes No
Video conferencing	Yes No	Yes No	Yes No
Voice over Internet (VoIP)	Yes No	Yes No	Yes No
Online dating and matrimonial services	Yes No	Yes No	Yes No
Online lectures	Yes No	Yes No	Yes No
Collaboration with schoolmates	Yes No	Yes No	Yes No
Accessing e-government services	Yes No	Yes No	Yes No

Appendix 4.14

Watching television/cable/satellite	Increased	No change	Decreased
Shopping in stores	Increased	No change	Decreased
Working at home	Increased	No change	Decreased
Reading newspapers/books/magazines	Increased	No change	Decreased
Working in the office	Increased	No change	Decreased
Commuting in traffic	Increased	No change	Decreased
Spending time with family	Increased	No change	Decreased
Spending time with friends	Increased	No change	Decreased
Attending social events	Increased	No change	Decreased
Time spent on sport	Increased	No change	Decreased
Time spent on hobbies	Increased	No change	Decreased
Time spent on sleeping	Increased	No change	Decreased
Time spent alone (doing nothing)	Increased	No change	Decreased
Studying	Increased	No change	Decreased
Household work	Increased	No change	Decreased
Receiving/ making phone calls	Increased	No change	Decreased
Doing charity and social works	Increased	No change	Decreased
Outdoor recreation (DIY, pet care)	Increased	No change	Decreased
Outdoor entertainment (concerts, cinema)	Increased	No change	Decreased
Visiting or meeting friends or relatives	Increased	No change	Decreased

Chapter 5

Development of
Survey Instrument:
Confirmatory Survey

Abstract

The previous chapter (Chapter 4) described the development and validation of a survey instrument for the purpose of data collection in order to examine broadband adoption, usage and impact. Chapter 3 provided a discussion and justification of the data collection and analysis methods. This chapter presents the findings obtained from a nationwide survey that was conducted to examine the adoption, usage and impact of broadband in UK households. The chapter is structured as follows. The next section presents a response rate of the survey and descriptions of how the non-response bias test was conducted. This is followed by a description of the reliability test conducted to assess the internal consistency of the survey instrument. The findings relating to the factor analysis are then presented. Following this, an overall discussion of the instrument is presented. Finally, a summary of the chapter is provided.

Response Rate and Non-Response Bias

A detailed discussion of the response rate estimation process is provided in Chapter 3. Therefore, in this chapter the estimation process is not described. The total response rate obtained in this research is 22.4 percent, which is considered as a good response rate within the field of IS research.

To test whether the characteristics of the respondents from the original responses are similar to the non-respondents, a *t*-test was conducted for the demographics (i.e., age and gender), Internet access at home, type of Internet connection at home, and all constructs from attitudinal, normative, and control categories. The findings are illustrated in Table 5.1. The *t*-test on demographics and all key constructs except the primary influence of the study showed no significant differences between the respondents and non-respondents (Table 5.1). There is a significant difference between the original responses and responses from the non-respondents for the primary influence construct. The possible explanation could be that the non-respondents were those who were more influenced by primary influence.

The majority of variables produced non-significant results in terms of non-response bias; this suggests that those non-respondents who returned the completed questionnaires after reminders were similar to the respondents from the original responses. Hence, this provides evidence that within the sample used for this research, it is unlikely that the findings were affected due to non-response bias.

Table 5.1. t-test to examine non-response bias

Variables	t	df	p
Age	.766	355	.444
Gender	.557	353	.578
Internet access at home	.646	356	.519
Type of connection	-1.609	306	.109
BI	-.547	356	.585
RA	.377	356	.707
UO	-.996	356	.320
HO	.845	356	.398
SQ	.161	306	.872
PI	-2.271	356	.024
SI	-.834	356	.405
K	.520	356	.604
SE	.072	356	.942
FCR	-1.079	356	.281

The research instrument was tested for its reliability, construct validity, and method bias (i.e., effect of question ordering). The subsequent sections illustrate reliability, show the construct validity, and present the computed values that demonstrate the absence of a method bias.

Reliability Test

Table 5.2 illustrates the Cronbach's coefficient alpha values that were estimated to examine the internal consistency of the measures. Cronbach's α varied between 0.91 for the utilitarian construct and 0.79 for both hedonic outcomes and service quality constructs. Both secondary influence and self-efficacy possessed a reliability value of 0.90. Cronbach's α for the remaining five constructs varied between 0.80 and 0.90. Two constructs—facilitating conditions resources and knowledge—had Cronbach's α at 0.81 and for relative advantage and primary influence there were values of alpha at 0.84. The dependent construct behavioural intention possessed an alpha of 0.87.

Hinton et al. (2004) have suggested four cut-off points for reliability, which includes excellent reliability (0.90 and above), high reliability (0.70-0.90), moderate reliability (0.50-0.70), and low reliability (0.50 and below) (Hinton et al., 2004, p. 364). These values suggest that of the 10 constructs, three possess excellent reliability and the

Table 5.2. Reliability of measurements

Constructs	N	Number of Items	Cronbach's Alpha (α)	Type
Behavioural Intention	358	2	.8790	High Reliability
BISP	**308**	1	---	---
Relative Advantage	358	4	.8481	High Reliability
Utilitarian Outcomes	358	10	.9131	**Excellent Reliability**
Hedonic Outcomes	358	4	.7968	High Reliability
Service Quality	**308**	4	.7912	High Reliability
Primary Influence	358	3	.8420	High Reliability
Secondary Influence	358	2	.9034	**Excellent Reliability**
Facilitating Conditions Resources	358	4	.8114	High Reliability
Knowledge	358	3	.8193	High Reliability
Self-efficacy	358	3	.9026	**Excellent Reliability**

Legend: BISP=Behavioural Intention to change subscriber N= Sample Size

remaining seven illustrate high reliability. None of the constructs demonstrated a moderate or low reliability (Table 5.2).

The high Cronbach's α values for all constructs imply that they are internally consistent. This means all items of each constructs are measuring the same content universe (i.e., construct). For example, both the items of BI are measuring the same content universe of behavioural intention. Similarly, all 10 items of UO are measuring the content universe of utilitarian outcomes construct. In brief, the higher the Cronbach's α value of a construct, the higher the reliability is of measuring the same construct.

Factor Analysis

In order to verify the construct validity (convergent and discriminant validity), a factor analysis was conducted utilising principal component analysis (PCA) with the Varimax rotation method. The results of the PCA are presented in Tables 5.3, 5.4, and 5.5.

Before conducting a factor analysis, it is essential to perform a test for sampling adequacy and sphericity. These two tests confirm whether it is worth proceeding with factor analysis (Hinton et al., 2004).

Kaiser-Meyer-Olkin Measure of Sampling Adequacy (KMO) Test and Bartlett's Test of Sphericity

The Kaiser-Meyer-Olkin (KMO) measure of sampling adequacy was first computed to determine the suitability of employing factor analysis, and the results are presented in Table 5.3. The KMO is estimated using correlations and partial correlations in order to test whether the variables in a given sample are adequate to correlate. A general 'rule of thumb' is that as a measure of factorability, a KMO value of 0.5 is poor, 0.6 is acceptable and a value closer to 1 is better (Brace et al., 2003; Hinton et al., 2004).

The results illustrated in Table 5.3 suggest that the KMO is well above the recommended acceptable level of 0.6 as the obtained value is 0.85. These results confirm that the KMO test supports the sampling adequacy and it is worth conducting a factor analysis. This means that higher KMO values indicate the possibility of factor existence in data as it was assumed in the conceptual model.

Bartlett's test of sphericity is conducted for the purpose of confirming the relationship between the variables. If there is no relationship, then it is irrelevant to undertake factor analysis. As a general rule, a p value <0.05 indicates that it is appropriate

Table 5.3. KMO and Bartlett's test

Kaiser-Meyer-Olkin Measure of Sampling Adequacy.		.858
Bartlett's Test of Sphericity	Approx. Chi-Square	6033.447
	Df	666
	Sig.	.000

to continue with the factor analysis (Brace et al., 2003; Hinton et al., 2004). The results illustrated in Table 5.3 suggest that the calculated *p* value is < 0.001, which means that there are relationships between the constructs in question. Therefore, it was considered appropriate to continue with the factor analysis.

Eigenvalues

As mentioned prior, factor analysis was conducted utilising principal component analysis as an extraction method and Varimax with Kaiser normalisation as a rotation method. Table 5.4 summarises the eigenvalues and explained total variance for the extracted components.

According to a general 'rule of thumb,' only those factors with eigenvalues greater than 1 should be considered important for analysis purposes (Hinton et al., 2004; Straub et al., 2004). The results presented in Table 5.4 suggest that all nine constructs included in the factor analysis possess eigenvalues greater than 1. Results from the analysis also suggest that no extracted new factor consisted of eigenvalues greater than 1.

Factor Loadings

The rotated component matrix presented in Table 5.5 shows the factor loadings for all nine constructs. The statistics presented in Table 5.4 clearly suggest that the nine components loaded. All the items loaded above 0.40, which is the minimum recommended value in IS research (Straub et al., 2004). Also, cross loading of the items was not found above 0.40.

All 10 items of the utilitarian outcomes construct loaded on component 1. Therefore, the first component represents the underlying constructs of utilitarian outcomes. For this construct, coefficients varied between 0.51 and 0.78. All four items of the facilitating conditions resources construct loaded on component 2. Therefore, the second component represents the underlying constructs of facilitating conditions

Table 5.4. Eigenvalues and total variance explained

C	Initial Eigenvalues			Extraction Sums of Squared Loadings			Rotation Sums of Squared Loadings		
	Total	percent of V	Cumulative percent	Total	percent of V	Cumulative percent	Total	percent of V	Cumulative percent
1	9.766	26.395	26.395	9.766	26.395	26.395	5.358	14.482	14.482
2	3.324	8.984	35.379	3.324	8.984	35.379	2.677	7.236	21.718
3	2.551	6.894	42.273	2.551	6.894	42.273	2.675	7.229	28.947
4	2.189	5.916	48.189	2.189	5.916	48.189	2.616	7.071	36.018
5	1.632	4.411	52.600	1.632	4.411	52.600	2.568	6.940	42.958
6	1.536	4.151	56.751	1.536	4.151	56.751	2.540	6.865	49.823
7	1.458	3.941	60.692	1.458	3.941	60.692	2.354	6.361	56.184
8	1.290	3.487	64.179	1.290	3.487	64.179	2.188	5.915	62.098
9	1.115	3.014	67.193	1.115	3.014	67.193	1.885	5.095	67.193

Extraction Method: Principal Component Analysis; **Legend:** C = Components; percent of V= Percentage of Variance

resources. The coefficient for this extracted component varies between 0.64 and 0.78. All four items of the service quality construct loaded on component 3. Therefore, the third component represents the underlying constructs of service quality. Coefficients for this component varied between 0.65 and 0.85. All four of the relative advantage related items loaded on the fourth component and loadings for this component varied between 0.58 and 0.72; this confirms that the fourth component represents the underlying constructs of relative advantage (Table 5.5).

All three items of the control construct self-efficacy loaded on component 5 with loadings that vary between 0.77 and 0.84. Thus, the fifth component represents the underlying constructs of self-efficacy. All four items related to the hedonic outcomes were loaded on component 6. The coefficients values for this component range between 0.60 and 0.85, and so the sixth component represents the underlying constructs of hedonic outcomes. All three items related to the primary influence construct loaded on the seventh component. The coefficients value was obtained from 0.65 and 0.89. This means that the seventh component represents the primary influence construct (Table 5.5).

The three items related to the control construct knowledge were loaded on the eighth component and the loadings range between 0.61 and 0.75. This means that the eighth component represents the underlying constructs of hedonic outcomes (Table 5.5). Finally, all items related with the secondary influence construct loaded on the ninth component. This construct comprised only two items and the coefficients of these two items were 0.91 and 0.90. Therefore, the ninth component represents the

Table 5.5. Rotated component matrix

Items	Component								
	1(UO)	2 (FCR)	3 (SQ)	4 (RA)	5 (SE)	6 (HO)	7 (PI)	8 (K)	9 (SI)
UO1	**.788**	.094	.102	.041	.070	.025	.087	-.021	-.069
UO6	**.783**	.116	.060	.095	.086	.057	-.016	.136	-.051
UO8	**.758**	.106	.054	.053	.035	.062	.121	-.038	.093
UO5	**.740**	.079	.025	.041	.070	.146	-.034	.077	.118
UO4	**.682**	.121	.094	.194	.027	.084	.041	.153	.014
UO2	**.679**	.041	.028	.107	.019	-.071	.079	.174	.168
UO3	**.663**	.124	.035	.188	.261	.096	.132	.028	-.063
UO10	**.564**	.240	.179	.347	.143	.100	.227	.184	-.006
UO7	**.520**	-.002	-.078	.368	.169	-.038	.134	.109	.021
UO9	**.519**	.309	.139	.246	.058	.021	.227	.227	.096
FCR3	.171	**.780**	.190	.079	.130	.103	.024	.022	.020
FCR1	.133	**.768**	.133	.073	.228	-.037	.084	.201	.020
FCR4	.107	**.687**	.029	.056	.238	-.026	.067	.212	.102
FCR2	.234	**.649**	.016	.206	-.089	.058	.042	-.016	-.070
SQ4	.087	.068	**.858**	.053	.134	.017	.041	.014	-.048
SQ1	.057	.132	**.794**	.063	-.083	.024	.011	.024	.007
SQ3	.041	.017	**.769**	.013	.225	.068	.183	.094	.032
SQ2	.111	.081	**.650**	-.047	-.027	.089	.097	.038	.102
RA4	.137	.153	.038	**.728**	.125	.022	.001	-.073	.105
RA2	.197	.124	.053	**.706**	.129	.118	.017	.308	.022
RA1	.222	.026	-.039	**.683**	.165	-.054	.048	.256	.017
RA3	.373	.175	.054	**.589**	.112	.022	-.013	.196	-.080
S2	.120	.115	.095	.117	**.844**	.055	-.035	.178	.003
S3	.230	.188	.088	.179	**.795**	.013	.016	.262	.032
S1	.172	.183	.045	.241	**.771**	-.005	.095	.139	-.023
HO2	.048	.010	.024	.135	.099	**.853**	.111	-.063	-.011
HO3	.081	.030	.109	-.151	-.049	**.793**	.133	.099	.135
HO1	.226	.090	.038	.116	.148	**.767**	.187	-.057	.047
HO4	-.028	-.013	.060	-.015	-.138	**.600**	-.055	.208	.268
PI1	.092	-.010	.093	.036	.005	.132	**.897**	.063	.108
PI2	.123	.067	.065	-.017	.058	.163	**.864**	.015	.117

Continued on following page

Table 5.5 continued

PI3	.298	.216	.271	.077	-.008	.064	**.654**	.033	.036
K3	.215	.123	.115	.255	.192	.054	.011	**.758**	-.021
K2	.182	.064	.067	.163	.311	.020	-.025	**.754**	-.010
K1	.173	.314	.015	.147	.130	.108	.177	**.615**	.033
SI1	.065	.034	.034	.112	.011	.143	.160	-.049	**.910**
SI2	.108	.026	.061	-.021	.005	.193	.091	.039	**.903**

Extraction Method: Principal Component Analysis.
Rotation Method: Varimax with Kaiser Normalisation.

secondary influence construct. There were no cross loading above 0.40 for any of the nine components (Table 5.5).

The factor analysis results satisfied the criteria of construct validity including both the discriminant validity (loading of at least 0.40, no cross-loading of items above 0.40) and convergent validity (eigenvalues of 1, loading of at least 0.40, items that load on posited constructs) (Straub et al., 2004). This confirms the existence of the construct validity (both discriminant validity and convergent validity) in the instrument measures of this research that were utilised for data collection (Table 5.5). This means that the collected data and findings that were obtained from this instrument are reliable.

Stevens (1996) provided the following three recommendations regarding the reliable factors. First, components with four or more loadings above 0.60 in absolute value are reliable, regardless of the sample size. Second, components with about 10 or more with 0.40 loadings are reliable as long as the sample size is greater than about 150. Third, components with only a few loadings should not be interpreted unless the sample size is at least 300 (Stevens, 1996). The results that are illustrated in Table 5.5 and presented previously satisfied all the three criteria recommended by Stevens (1996). Therefore, it confirms that the extracted components are reliable and that construct validity exists (Table 5.5).

Total Variance Explained

Table 5.4 summarises the explained total variance for the extracted components that are shown in Table 5.5. All constructs had eigenvalues greater than 1 and in combination, accounted for a total of 67.13 percent variance in data. Variance contributed by each construct varies before and after rotation. Values presented hereafter represent before-rotation variance and after-rotation values and are illustrated in Table 5.4.

Within this category, the maximum variance of 26.39 percent was explained by the utilitarian outcomes construct. Amongst the attitudinal constructs, service quality had the second largest variance in data (6.89 percent). The relative advantage construct followed this with a variance of 5.91 percent. The hedonic outcomes contribute to a variance of 4.15 percent (Table 5.4).

The minimum variation of 3.01 percent was accounted for by the normative construct 'secondary influence'. The other normative construct (primary influence) accounted for only a 3.94 percent variance in data (Table 5.4).

The first control construct, self-efficacy, accounted for a total variance of 4.44 percent. The second control construct, knowledge, accounted for a total of 3.48 percent variance. The third control construct, facilitating conditions resources, accounted for 8.98 percent variance in the data (Table 5.4).

Findings from both the reliability test and factor analysis, which respectively confirms internal consistency of measures and construct validities (i.e., convergent and discriminant validity), suggest that it is appropriate to create aggregated measures by averaging the means of all items of each construct. Chapter 3 has already provided a discussion on this issue.

Test for Ordering of Questionnaire Items

Straub et al. (2004) argued that as a result of the lack of randomisation of items for a particular construct, respondents may sense the inherent constructs via the ordering of questionnaire items and therefore their response may introduce a bias, which is termed as a methods bias. This type of bias is considered to be a threat to construct validity (Straub, 2004). To examine if any method bias exists within this study, a t-test was conducted for two samples, one with randomisation of questionnaire items and one without it. Table 5.6 illustrates the results that showed no significant difference between the obtained responses from the randomised and non-randomised questionnaire. Therefore, it is unlikely that a method bias exists in the collected data, or more specifically, that the questionnaire items ordering in this particular instance contributed to the pattern of responses; instead the findings presented the 'true scores.' In brief, there is no threat to the construct validity due to a method bias in the data.

Table 5.6. t -Test to compare means of aggregated measures obtained from randomised (N= 40) and non-randomised (N=318) questionnaire

	t Value	Df	*p*
BI	.209	356	.835
RA	.122	356	.903
UO	.089	356	.930
HO	.745	356	.457
SQ	-1.399	306	.163
PI	-.334	356	.739
SI	-.533	356	.594
K	-1.361	356	.174
SE	.910	356	.364
FCR	.269	356	.788

Discussions

Response Rate and Non-Response Bias

A 22.4 percent response rate was obtained in this research. Cornford and Smithson (1996) suggested that, within IS research, a response rate of 20 percent is considered to be acceptable and if the response rate is approximately 10 percent then that means the questionnaire design was poor (Cornford & Smithson, 1996). According to Fowler (2002), the majority of surveys produce response rates between two extreme values that are 5 percent at the lower end and 95 percent at the higher end, with response rates above 20 percent being considered as satisfactory (Fowler, 2002). Considering the two recommended levels (Cornford & Smithson, 1996; Fowler, 2002), the survey response of this research is considered to be satisfactory and acceptable.

However, despite the response rate, a non-response bias could arise in the findings. Therefore, it is essential to conduct a non-response bias test in order to demonstrate whether the non-respondents are similar to the respondents (Fowler, 2002; Karahanna et al., 1999). In this research a *t*-test was undertaken to determine whether the characteristics of the respondents from the original responses are similar to the non-respondents. The *t*-test was conducted for demographics (i.e., age and gender), Internet access at home, the type of Internet connection at home, and all constructs from the attitudinal, normative and control categories. The findings are illustrated in Table 5.1, which suggests that the demographics and all key constructs except the primary influence of the study showed no significant differences between the respondents and non-respondents. There is a significant difference between the original responses and responses from the non-respondents for the primary influence construct. Since primary influence is the only construct of the 14 variables that was tested for the response bias and was found to be significant, it is possible that the non-respondents were those who were more influenced by primary influence. This suggests that those non-respondents who returned the completed questionnaire after reminders were similar to the respondents from the original responses. Hence, this provides evidence that within the sample used for this research, there are minimal chances that it is likely that the data has a non-response bias.

Recent studies have demonstrated that a low response rate in survey research does not necessarily produce a non-response bias (Keeter et al., 2000; Karahanna et al., 1999). Keeter et al. (2000) compared two telephone surveys, in which the first one obtained a 36 percent response and the second had a 60 percent response rate. This study reported that both the surveys with a different response rate provided similar results with minor statistically significant differences (Fowler, 2002; Keeter et al., 2000). In a study that employed the postal questionnaire as a data collection tool, it was found that there were no significant differences between the responses of respondents and non-respondents for both the demographic variables and key constructs (Karahanna et al., 1999). Since this research reported similar findings as the previous two studies, it can be said that it is less likely that these findings are affected by a non-response bias.

Instrument Validation

To establish and demonstrate rigour in the findings of positivist research, validity should be undertaken both prior to and after final data collection (Straub et al., 2004). The validation process suggested for application to the cases is one where research either utilises previously validated instruments or creates new instruments (Straub et al., 2004, p. 412). Although application of validation is recommended in both situations, it is essential in the latter case where a study employs newly created instruments for data collection (Straub et al., 2004, p. 414). Since this study created a new research instrument for examining broadband adoption, usage and impact, the utmost care was taken to validate the newly created instrument. This section provides an overall picture of the validation process and also briefly discusses if the undertaken validity measures and their outcomes are on a par with the recommendations made in IS research.

The recommended validities include content validity, construct validity, reliability, manipulation validity, and the common method bias (Straub et al., 2004). Amongst these validities, this research examined all the suggestions except for manipulation validity. Manipulation validity, that forms an essential component of experimental

Figure 5.1. Development and validation process of research instrument

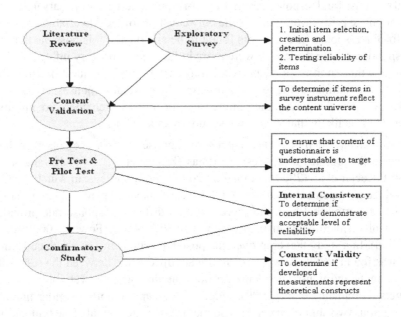

research, was not employed in this research as it was suggested to be inappropriate in the context of survey research (Straub et al., 2004). Figure 5.1 depicts the overall process of creating and validating a new research instrument. The justification for undertaking each stage is provided in the previous chapter (Chapter 4) and the purpose is briefly illustrated in Figure 5.1.

The stages involved in the validation process comprised an exploratory survey, content validation, pre- and pilot- tests and finally the confirmatory study. Validities that are exercised in this research included content validity, reliability and construct validity (Figure 5.1 and Table 5.7). The previous study on broadband adoption in the South Korean context, although it employed post data collection validity measures such as reliability and construct validity, lacked the application of validity prior to data collection such as content validity. In order to create a reliable survey instrument and confidence in the research findings, this study employed both pre (i.e., content validity) and post data collection validities (i.e., reliability and construct validity) that are recommended in IS research (Straub et al., 2004).

Table 5.7 compares the validities that are undertaken in this research with the recommended standard in IS research. The first validity that was applied in this research was content validity, which utilised a quantitative approach (Lawshe, 1975) that is recommended by Straub et al. (2004). Since content validity was not applied in any of the previous studies on broadband adoption and diffusion, it was not possible to

Table 5.7. Summary of instrument validation process (Source: Straub et al., 2004)

Validity Component	Type	Technique Suggested	Heuristics	Technique Applied
				Remarks
Content	Highly Recommended	Literature review; expert panels or judges; content validity ratios [Lawshe, 1975]; Q-sorting	Items only included if Content Validity Ratios (CVR)=> 0.5	Content Validity Ratio (CVR)
				12 Experts rated items; Items that possess CVR <0.50 were dropped [Lawshe, 1975]
				This suggest that content validity exist in final instrument
Construct				
1. Discriminant Validity	Mandatory	MTMM; PCA; CFA; PLS AVE; Q-sorting	Latent Root Criterion (eigenvalue) of or above 1, Loadings of at least .40 (although some references suggest a higher cutoff); no crossloading of items above .40. Items that do not load properly may be dropped from the instrument [Churchill, 1979].	Principal Component Analysis (PCA)
				1. Latent Root Criterion (eigenvalue) of 1
				2. All items loaded above .40;
				3. No crossloading of items above .40.
				This confirms existence of discriminant validity in final instrument.
2. Convergent Validity	Mandatory	MTMM; PCA; CFA; Q-sorting	Eigenvalues of 1; loadings of at least .40; items load on posited constructs; items that do not load properly are dropped.	Principal Component Analysis (PCA)
				1. Eigenvalues of 1
				2. All loadings >.40
				3. All items loaded on posited constructs;
				4. No items needed to be dropped due to crossloading
				This confirms existence of convergent validity in final instrument.

continued on following page

Table 5.5. continued

		MTMM, CFA through LISREL	Collect data at more than one period; collect data using more than one method; separate data collection of IVs from DVs	*t*-Test
3. Methods bias	Highly Recommended			*p* value for all constructs found to be non significant that suggested there was no bias due to items ordering in questionnaire
Reliability	Mandatory	Cronbach's α; correlations; SEM reliability coefficients	Cronbach's α should be above .60 for exploratory, .70 for confirmatory; in PLS, should be above .70; in LISREL, EQS, and AMOS, should also be above .70.	Cronbach's α
Internal consistency				Cronbach's α for all constructs found above .79, which suggest that internal consistency exist in instrument

compare this research's process and findings. Lawshe (1975) suggested that constructs that should be included in a study are those that account for a content validity ratio (CVR) of more than 0.5 on a 0 to 1 scale. The findings presented in Table 4.3 illustrate that all 11 constructs had a CVR value above 0.5; it can therefore be said that the instrument possessed an appropriate level of content validity. This means that the items in this instrument reflect the content universe to which the instrument of this research will be generalised (Straub et al., 2004).

Table 5.7 further illustrates that this research also undertook a construct validity and reliability test. Construct validity was performed utilising the PCA. Oh et al. (2003) also employed the PCA to confirm construct validity in a previous broadband adoption study. The standard recommendation (Straub et al., 2004) suggested that items should not be cross loaded over 0.40, but, Oh et al. (2003) suppressed the value below 0.50. Therefore, in this study it was not possible to consider whether any items cross loaded on any other constructs, and so it created a sense of doubt as to whether the construct validity existed in Oh et al. (2003). This study did not suppress values and Table 5.7 clearly demonstrates that this study meets the standard criteria (Straub et al., 2004) of all types of validities, namely convergent validity, discriminant validity and method bias (Table 5.7). This implies that the validated instrument provides an effective measure of the theoretical constructs included in the conceptual model.

Finally, the internal consistency of measures was assessed utilising a reliability test (i.e., Cronbach's α). Straub et al. (2004) suggested that, for a confirmatory study, reliability should be equal to or above 0.70 (Table 5.7). The reliability values reported in Oh et al. (2003) varies between 0.70 and 0.89 for various constructs. Reliability or the Cronbach's α value of various constructs in this research varies between 0.79 and 0.91, which means that all the constructs possessed reliability values above the minimum recommended level (Table 5.2). This suggests that measures of this study demonstrate an appropriate level of internal consistency.

Summary

This chapter presented the findings obtained from the data analysis of the confirmatory survey that was conducted to examine consumer adoption and usage and the impact of broadband in UK households. The findings were presented in several sections. The first step was to calculate the response rate of the survey and conduct a response bias test. The estimated response rate was 22.4 percent and the response bias test suggested that there was no significant difference for the demographic characteristics such as the age and gender of the respondents and non-respondents. Also, a response bias test showed no significant differences between the responses of respondents and non-respondents with regards to key constructs such as relative advantage, utilitarian outcome, hedonic outcome, service quality, secondary influence, knowledge, self-efficacy, and facilitating conditions resources.

Following the non-response bias test, there was a discussion of the validation and findings obtained on the adoption of broadband. The section initially presented findings that illustrated the reliability test, construct validity, and effect of the question ordering. The reliability test confirmed that the measures are internally consistent, as all the constructs possessed a Cronbach's alpha above 0.70.

The construct validity was established utilising the PCA. The results of the PCA provided evidence of higher KMO values (0.858), a significant probability of Bartlett's test of sphericity (< .001), extraction of components consistent with the number of independent factors in the conceptual model (all the nine factors possessed eigenvalues above 1), factors which all loaded above 0.40 and no cross loading above 0.40. This confirmed that both types of the construct validity (i.e., convergent and discriminant) existed in the survey instrument.

A t-test was conducted to confirm if any difference occurred due to the ordering of questionnaire items. The results indicated no significant differences between the responses with or without the ordering of questionnaire items. This further strengthened the existence of the construct validity in the survey instrument.

This chapter also discussed and reflected upon the findings from the theoretical perspective. First, this chapter discussed the obtained response rate and the effects of the non-response bias. The discussion suggests that the response rate obtained within this study is satisfactory and falls within an acceptable range in IS research. Furthermore, the effect of the non-response bias was found to be minimal in this study, which means that the findings of this study are least likely to be affected by non-responses.

Second, this chapter compared the outcomes of the instrument development and validation processes with a standard recommended within IS research in terms of content validity, construct validity and reliability. The comparison led to the conclusion that the research instrument possessed an appropriate level of content validity, reliability and construct validity and satisfied the standard criteria within IS research.

References

Brace, N., Kemp, R., & Snelgar, R. (2003). *SPSS for psychologists: A guide to data analysis using SPSS for windows.* New York: Palgrave Macmillan.

Churchill, G. A. Jr. (1979). A paradigm for developing better measures of marketing constructs. *Journal of Marketing Research, 16*(1), 64-73.

Cornford, T., & Smithson, S. (1996). *Project research in information systems: A student's guide.* London: Macmillan Press Ltd.

Fowler, F. J. Jr. (2002). *Survey research methods.* London: Sage Publications Inc.

Hinton, P. R., Brownlow, C., McMurray, I., & Cozens, B. (2004). *SPSS explained.* East Sussex, UK: Routledge Inc.

Karahanna, E., Straub, D. W., & Chervany, N. L. (1999). Information technology adoption across time: A cross-sectional comparison of pre-adoption and post-adoption beliefs. *MIS Quarterly, 23*(2), 183-213.

Keeter, S., Miller, C., Kohut, A., Groves, R. M., & Presser, S. (2000). Consequences of reducing nonresponse in a national telephone survey. *Public Opinion Quarterly, 64*(2), 125-148.

Lawshe, C. H. (1975). A quantitative approach to content validity. *Personnel Psychology, 28,* 563-575.

Oh, S., Ahn, J., & Kim, B. (2003). Adoption of broadband Internet in Korea: the role of experience in building attitude. *Journal of Information Technology, 18*(4), 267-280.

Straub, D. W., Boudreau, M-C, & Gefen, D. (2004). Validation guidelines for IS positivist research. *Communications of the Association for Information Systems, 13,* 380-427.

Stevens, J. (1996). *Applied multivariate statistics for the social sciences.* NJ: Lawrence Erlbaum Associates, Inc.

Section 1.3

The Empirical
Underpinning

Chapter 6

Empirical Findings:
Adoption, Usage, and Impact of Broadband

Abstract

The previous chapters (Chapters 4 and 5) described the development and validation of a survey instrument for the purpose of data collection in order to examine broadband adoption, usage, and impact. Chapter 3 provided a discussion and justification of the data collection and analysis methods. This chapter presents the findings obtained from the survey that was conducted to examine the adoption, usage, and impact of broadband in UK households. The chapter is structured as follows. The next section describes the demographic profiles of the survey respondents. This is followed by a description of the findings relating to the adoption of broadband. The findings relating to the usage of broadband are then presented. This is followed by a description of the effects of broadband usage on consumers'time allocation patterns in various daily life activities. Finally, the summary of the chapter is provided.

Respondents' Profile

A profile of the survey respondents is presented in Table 6.1. Of the 358 received responses, 26.1 percent of the respondents belonged to the 25-34 years age group, which formed the largest response category. The 35-44 years age group follows this with 21.6 percent. The least responsive category was the 65 years and above with 3.9 percent. In terms of gender, only 2 percent more responses were obtained from males (51 percent) in comparison to female (49 percent) respondents (Table 6.1).

The majority of the respondents possessed educational qualifications, with 34.6 percent having gained an undergraduate degree and 29.3 percent educated to postgraduate level. The least responsive educational category was the GNVQ/Diploma with an 8.8 percent response rate. As noted in Table 6.1, 11.7 percent of the respondents possessed GCSE level education and 15.5 percent possessed 'A' level education.

The occupational category with the highest amount of respondents was 'E,' which consisted of students, casual workers, and pensioners (37.1 percent). This was followed by category 'B,' which consisted of managers, teachers, and computer programmers (28.4 percent). C1 represented the third largest occupational category with 19.7 percent responses, followed by category 'A' with 11.0 percent response. The least responsive occupational categories were 'D' and 'C2,' with response rates of 1.7 percent and 2.0 percent respectively (Table 6.1). The occupational categories, as described in Chapter 2, were derived from the marketing literature where mainstream professionals, such as doctors, lawyers, and judges with the responsibility of more then 25 staff are classified as occupational category 'A' (Gilligan & Wilson, 2003; Rice 1997). Occupations wherein individuals had a responsibility of less then 25 staff and academics are grouped as social grade 'B.' Skilled non-manual workers fall within the occupational category 'C1' and 'C2.' Unskilled manual workers belong to occupational category 'D.' Finally, housewives, retired individuals, students, and unemployed citizens were placed in category 'E' (Gilligan & Wilson, 2003; Rice 1997). As broadband provides a function to students and unemployed people who are engaged in job hunting, these groups are more likely to adopt broadband, although they belong to the lower occupation category 'E.'

Responses for the household income categories varied between a response rate of 17.9 percent for the £20-29 K and 7.2 percent for £50-59K category. The least annual household income group (<=£10K) was represented by a 9.5 percent response, whilst the largest income group (=>£70K) was represented with a 10.4 percent response rate (Table 6.1).

Of the 358 respondents, 308 (86 percent) had Internet access at home and 50 (14 percent) did not. Of the 308 (86 percent) respondents who possessed Internet access at home, 101 (32.8 percent) had a narrowband connection and the remaining 207 (67.2 percent) respondents had a broadband connection (Table 6.1).

Table 6.1. Profile of survey participants

	Categories	Frequency	Percent
	<=24	75	21.0
	25-34	93	26.1
	35-44	77	21.6
Age	45-54	68	19.0
	55-64	30	8.4
	=>65	14	3.9
	Total	**357**	**100.0**
	Male	181	51.0
Gender	Female	174	49.0
	Total	**355**	**100.0**
	GCSC	40	11.7
	GNQV/Diploma	30	8.8
	A level	53	15.5
Education	UG	118	34.6
	PG	100	29.3
	Total	**341**	**100.0**
	A	38	11.0
	B	98	28.4
	C1	68	19.7
Occupation	C2	7	2.0
	D	6	1.7
	E	128	37.1
	Total	**345**	**100.0**
	<10 K	33	9.5
	10-19 K	60	17.3
	20-29 K	62	17.9
	30-39 K	60	17.3
Income	40-49 K	38	11.0
	50-59 K	25	7.2
	60-69 K	33	9.5
	=> 70 K	36	10.4
	Total	**347**	**100.0**

Continued on following page

Table 6.1. continued

Internet access at home	Yes	308	86.0
	No	50	14.0
	Total	**358**	**100.0**
Type of internet access at home	Narrowband	101	32.8
	Broadband	207	67.2
	Total	**308**	**100.0**

Adoption of Broadband

Descriptive Statistics

Table 6.2 presents the means and standard deviations of the items related to all 11 constructs included in the study to measure the perceptions regarding broadband adoption. The means and standard deviations of aggregated measures for all the 11 constructs are also illustrated in Table 6.3.

The respondents showed strong agreement for both items of the behavioural intentions (BI1 and BI3), as the mean score varies between 5.78 (SD=1.62) and 6.05 (SD=1.48), (Table 6.2) with an average score of 5.92 (SD=1.55) (Table 6.3). Only the item (BI2) of the behavioural intention to change service provider (BISP) construct was less agreed by survey respondents (M = 3.42, SD = 1.88) (Table 6.2).

The respondents agreed strongly for all of the items of the relative advantage constructs, where item RA1 scored the maximum (M = 6.39, SD = 1.11) and minimum (M = 6.10, SD = 1.28) for item RA3 (Table 6.2) with the high average score of aggregate measure (M = 6.31, SD = 1.17) (Table 6.3). A strong agreement was also made for the utilitarian outcomes (M = 5.60, SD = 1.45) and service quality (M = 4.67, SD = 1.70) constructs by survey respondents (Table 6.3). The importance of hedonic outcomes was less agreed with an average mean score of 3.51 and standard deviations of 1.92 (Table 6.3).

Amongst the normative constructs, primary influence rated above average (*M* = 4.75, *SD* = 1.68) and was agreed more strongly than the secondary influence which was rated slightly above than average (*M* = 3.65, *SD* = 1.80) on a 7 point likert scale (Table 6.3). Self-efficacy was rated stronger (*M* = 6.24, *SD* = 1.29) than the other control constructs, namely knowledge (*M* = 5.59, *SD* = 1.49) and facilitating conditions resources (*M* = 4.67, *SD* = 1.70) (Table 6.3).

Table 6.2. Descriptive statistics

SN	Items	N	Mean	SD	SN	Items	N	Mean	SD
1	BI1	358	5.78	1.624	21	HO4	358	1.92	1.587
2	BI2	308	3.42	1.882	22	SQ1	308	4.74	1.859
3	BI3	358	6.05	1.481	23	SQ2	308	4.39	1.817
4	RA1	358	6.39	1.119	24	SQ3	308	4.47	1.593
5	RA2	358	6.38	1.131	25	SQ4	308	5.07	1.533
6	RA3	358	6.10	1.287	26	PI1	358	4.68	1.642
7	RA4	358	6.39	1.148	27	PI2	358	4.62	1.632
8	UO1	358	5.89	1.328	28	PI3	358	4.94	1.768
9	UO2	358	5.63	1.396	29	SI1	358	3.80	1.844
10	UO3	358	5.76	1.440	30	SI2	358	3.49	1.783
11	UO4	358	5.73	1.480	31	K1	358	5.44	1.540
12	UO5	358	5.32	1.550	32	K2	358	5.61	1.511
13	UO6	358	5.64	1.407	33	K3	358	5.73	1.417
14	UO7	358	5.57	1.341	34	S1	358	6.32	1.314
15	UO8	358	5.41	1.520	35	S2	358	6.17	1.308
16	UO9	358	5.26	1.601	36	S3	358	6.23	1.239
17	UO10	358	5.82	1.414	37	FCR1	358	5.57	1.720
18	HO1	358	4.71	2.040	38	FCR2	358	4.57	1.793
19	HO2	358	4.04	1.995	39	FCR3	358	4.79	1.809
20	HO3	358	3.35	2.050	40	FCR4	358	5.60	1.582

N: Total number of responses. *SD*: Standard Deviation

The descriptive statistics are the cumulative scores obtained from both broadband and narrowband consumers, and it is expected that the mean score may differ for the two groups. Hence, the findings that illustrate the cross sectional view are presented in the next subsection, which demonstrates that broadband consumers' perception of having broadband was significantly higher than its narrowband counterpart.

The Difference between Broadband Adopters and Non-Adopters

Table 6.4 presents the means and standard deviations of the 10 aggregate measures included in the study for both narrowband and broadband consumers. Table 6.4 also provides the results of the *t*-test, which tested the differences between the narrowband and broadband consumers on these constructs. The findings indicate

Table 6.3. Summary of descriptive statistics

SN	Construct	NI	N	Descriptive			
				Mean	Min	Max	*SD*
1	Behavioural Intention	2	358	5.92	5.78	6.05	1.55
2	BISP	1	308	3.42	---	---	1.88
3	Relative Advantage	4	358	6.31	6.09	6.39	1.17
4	Utilitarian Outcomes	10	358	5.60	5.25	5.89	1.45
5	Hedonic Outcomes	4	358	3.51	1.92	4.71	1.92
6	Service Quality	4	308	4.67	4.39	5.06	1.70
7	Primary Influence	3	358	4.75	4.62	4.94	1.68
8	Secondary Influence	2	358	3.65	3.49	3.79	1.81
9	Facilitating Conditions Resources	4	358	5.13	4.57	5.60	1.73
10	Knowledge	3	358	5.59	5.44	5.73	1.49
11	Self-efficacy	3	358	6.24	6.17	6.32	1.29

NI: Total number of variables or items. *N:* Total number of responses. *SD:* Standard Deviation

that with the exception of secondary influence, the narrowband and broadband consumers differ significantly on the mean score for the remaining nine constructs. Even though both groups (i.e., narrowband and broadband consumers) view the adoption of broadband positively overall, the mean scores indicate that broadband consumers have significantly more positive perceptions on the various constructs than narrowband consumers.

Discriminant Analysis

To confirm the effectiveness of various factors for discriminating adopters from non-adopters, a discriminant analysis was performed using broadband adoption as the dependent variable and behavioural intention, relative advantage, utilitarian outcomes, hedonic outcomes, service quality, primary influence, secondary influence, facilitating conditions resources, knowledge, and self-efficacy as the predictor variables. A total of 308 cases were analysed. The findings are presented in Tables 6.5-6.6. The univariate ANOVAs revealed that the narrowband and broadband consumers differed significantly on all the predictor variables except for secondary influence (Table 6.5). A single determinant function was calculated (Table 6.5). The value of this function was significantly different for the narrowband and broadband consumers ($\chi^2 (10, N = 308) = 128.867, p < .001$). The correlations between the predictor variables and the discriminant function suggest that behavioural intention

Table 6.4. t-tests to examine equality of group means

Construct	Type of connection	N	M	M Difference	SD	t	df	p (2-tailed)
BI	Narrowband	101	5.18		1.74			
	Broadband	207	6.54	1.36	.71	9.70	306	.000
	Broadband	207	3.46		1.90			
RA	Narrowband	101	6.14		.95			
	Broadband	207	6.53	.38	.67	4.10	306	.000
UO	Narrowband	101	5.17		1.12			
	Broadband	207	5.92	.74	.83	6.56	306	.000
HO	Narrowband	101	3.21		1.53			
	Broadband	207	3.62	.40	1.46	2.24	306	.025
SQ	Narrowband	101	4.02		1.34			
	Broadband	207	4.97	.94	1.22	6.18	306	.000
PI	Narrowband	101	4.43		1.52			
	Broadband	207	5.02	.59	1.36	3.44	306	.001
SI	Narrowband	101	3.74		1.74			
	Broadband	207	3.57	-.17	1.73	-.819	306	.414
K	Narrowband	101	5.36		1.29			
	Broadband	207	5.85	.49	1.08	3.50	306	.001
SE	Narrowband	101	6.04		1.28			
	Broadband	207	6.44	.39	.93	3.05	306	.003
FCR	Narrowband	101	4.68		1.49			
	Broadband	207	5.62	.94	1.04	6.41	306	.000

was the best predictor of the future adoption of broadband whilst secondary influence was found to be the least useful (Table 6.5).

Overall, the discriminant function successfully predicted the outcome for 80.2 percent of the cases, with accurate predictions being made for 72.3 percent of the narrowband consumers and 84.1 percent of the broadband consumers (Table 6.6).

Demographic Differences

Age and Adoption of Broadband

Table 6.7 illustrates that the adoption of broadband amongst consumers increases with age; however, the subscription rate fell after the 54 years range and only 1

Table 6.5. Tests of equality of group means and structure matrix

	Tests of Equality of Group Means				Structure Matrix	
	F	df1	df2	p		Function
BI	94.154	1	306	.000	BI	.759
RA	16.833	1	306	.000	UO	.513
UO	43.083	1	306	.000	FCR	.502
HO	5.056	1	306	.025	SQ	.484
SQ	38.263	1	306	.000	RA	.321
PI	11.849	1	306	.001	K	.274
SI	.670	1	306	.414	PI	.269
K	12.310	1	306	.001	SE	.237
SE	9.211	1	306	.003	HO	.176
FCR	41.192	1	306	.000	SI	-.064

Table 6.6. Classification results [a]

	Type of connection	Predicted Group Membership		Total
		Narrowband	Broadband	
Count	Narrowband	73	28	101
	Broadband	33	174	207
%	Narrowband	72.3	27.7	100.0
	Broadband	15.9	84.1	100.0

[a.] 80.2% of original grouped cases correctly classified.

percent of subscribers were reported at the above 65 years category. The majority of broadband subscribers were between 25 and 54 years. The findings in Table 6.7 suggest that broadband consumers belong to the younger and middle-aged aged groups; however, the older age groups consisted of a majority of non-adopters. Pearson's chi-square test (Table 6.7) confirmed that there was a difference between the ages of the adopters and non-adopters of broadband (χ^2 (5, $N = 357$) = 15.016, $p = .010$).

A binary correlation test was also conducted to examine if there was any association between the age of respondents and broadband adoption. Table 6.8 presents the results obtained from this test. The findings suggest that there was a significant negative correlation between the age of respondents and broadband adoption (Table 6.8).

Table 6.7. Age as a determinant of broadband adopters and non-adopters

Age Categories	Non-adopters		Broadband adopters	
	Frequency	Percent	Frequency	Percent
Less than 24	25	16.7	50	24.2
25-34	43	28.7	50	24.2
35-44	28	18.7	49	23.7
45-54	29	18.7	39	18.8
55-64	13	8.7	17	8.2
More than 65	12	8	2	1
Total	150	100	207	100

Age X broadband adoption	χ^2Test (N=357)		
	Value	df	p (2-sided)
Pearson χ^2	15.016	5	.010

Table 6.8. Spearman's rho correlations to show association between age and broadband adoption

		Broadband Adoption
Age of Respondents	Correlation Coefficient	-.153(**)
**Correlation is significant at the 0.05 level (1-tailed).	Sig. (1-tailed)	.004
	N	357

Gender and Adoption of Broadband

In terms of gender differences, Table 6.9 illustrates that amongst the broadband adopters there are more males (53.6 percent) compared to females (46.4 percent). In contrast, within the non-adopters, the females (52.7 percent) exceeded the males (47.3 percent). Although these figures suggest gender differences between the adopters and non-adopters, it is not large enough to suggest the occurrence of any significance (Table 6.9). Table 6.9 illustrates that there were no significant differences between the genders of broadband adopters and non-adopters (χ^2 (1, N = 355) = 1.382, p = .240).

Education and Adoption of Broadband

Table 6.10 illustrates the educational attainment of broadband adopters and non-adopters. The findings suggest that the majority of adopters are educated to an

Table 6.9. Gender as a determinant of broadband adopters and non-adopters

Gender	Non-adopters		Broadband adopters	
	Frequency	Percent	Frequency	Percent
Male	70	47.3	111	53.6
Female	78	52.7	96	46.4
Total	148	100	207	100
Gender X broadband adoption	χ^2Test (N=355)			
	Value	**df**	***p*** (2-sided)	
Pearson χ^2	1.382	1	.240	

Table 6.10. Education as a determinant of broadband adopters and non-adopters

Education level	Non-adopters		Broadband adopters	
	Frequency	Percent	Frequency	Percent
GCSE	30	21.3	10	5
GNVQ/DIPLOMA	15	10.6	15	7.5
A LEVEL	22	15.6	31	15.5
UG	40	28.4	78	39
PG	34	24.1	66	33
Total	141	100	200	100
Education X broadband adoption	χ^2Test (N=341)			
	Value	**df**	***p*** (2-sided)	
Pearson χ^2	24.532	4	< .001	

undergraduate degree level (39 percent) followed by respondents who had post-graduate level (33 percent) education. 15.5 percent adopters of broadband had A-level qualifications. A small number of the adopters (5 percent) had an education level of GCSE followed by GNVQ (7.5 percent). In comparison to the adopters, the majority of non-adopters were reported to have lower levels of education. The educational qualification of GCSE had the highest percentage of non-adopters; of the 40 respondents who had GCSE level education, 30 were non-adopters (Table 6.10). The Pearson's chi-square test validated that there was a significant difference between the education levels of the adopters and non-adopters of broadband ($\chi^2(4, N = 341) = 24.532, p < .001$) (Table 6.10).

Also, a binary correlation test was conducted to examine if there was any association between the education level of respondents and broadband adoption. Table

Table 6.11. Spearman's rho correlations to show association between education and broadband adoption

		Broadband Adoption
Education of Respondents	Correlation Coefficient	.208(**)
** Correlation is significant at the 0.01 level (1-tailed).	Sig. (1-tailed)	.000
	N	341

Table 6.12. Occupation as a determinant of broadband adopters and non-adopters

Occupation Categories	**Non-adopters**		**Broadband adopters**	
	Frequency	**Percent**	**Frequency**	**Percent**
A	12	8.4	26	12.9
B	30	21	68	33.7
C1	38	26.6	30	14.9
C2	4	2.8	3	1.5
D	5	3.5	1	.5
E	54	37.8	74	36.6
Total	143	100	202	100

χ^2 Test (N=345)			
Occupation X broadband adoption			
	Value	df	*p* (2-sided)
Pearson χ^2	17.181	5	.004

6.11 presents the results obtained from this test. The findings suggest that there was a significant positive correlation between the education level of respondents and broadband adoption (Table 6.11).

Occupation and Adoption of Broadband

Table 6.12 illustrates the occupational category for both the adopters and non-adopters. This suggests that a total of 38 respondents from occupational category 'A' provided responses. Of those 38, 26 respondents in this category were adopters and 12 were non-adopters. Similar trends were observed for occupational category 'B,' which consisted of 68 adopters compared to 30 non-adopters. Occupational category 'E' also consisted of more adopters (74) than non-adopters (54). Contrary to this, occupational categories C1, C2 and D had more non-adopters than adopters (Table 6.12). The findings from Pearson's chi-square test also validated that there was a significant difference between the occupational categories of adopters and non-adopters of broadband (χ^2 (5, $N = 345$) = 17.181, $p = .004$) (Table 6.12).

Household Annual Income and Adoption of Broadband

The findings illustrated in Table 6.13 suggest that the minimum numbers (9.4 percent) of adopters belonged to the category with less then a £10 K annual household income. The second lowest income group, that is £10-19 K, had more non-adopters (26.9 percent) than adopters (10.4 percent). However, all the income categories above £10-19 K had more adopters than non-adopters (Table 6.13). Generally, the adopters exceeded the non-adopters in all the higher income levels. The Pearson's chi-square test confirmed that there was a significant difference between the household annual income category of the adopters and non-adopters of broadband (χ^2 (7, $N = 347$) = 28.401, $p < 0.001$) (Table 6.13).

A binary correlation test was also conducted to examine if there was an association between households' annual income and broadband adoption. The results obtained

Table 6.13. Household annual income as a determinant of broadband adopters and non-adopters

Income Categories	Non-adopters		Broadband adopters	
	Frequency	Percent	Frequency	Percent
Less than 10 K	14	9.7	19	9.4
10-19 K	39	26.9	21	10.4
20-29 K	28	19.3	34	16.8
30-39 K	29	20	31	15.3
40-49 K	10	6.9	28	13.9
50-59 K	5	3.4	20	9.9
60-69 K	11	7.6	22	10.9
More than 70 K	9	6.2	27	13.4
Total	145	100	202	100
χ^2 Test (N=347)				
Income X broadband adoption				
	Value	df	p (2-sided)	
Pearson χ^2	28.401	7	< .001	

Table 6.14. Spearman's rho correlations to show association between income and broadband adoption

		Broadband Adoption
Households Annual Income	Correlation Coefficient	.222(*)
* Correlation is significant at the 0.01 level (1-tailed).	Sig. (1-tailed)	.000
	N	347

from this test suggest that there was a significant positive correlation between household annual income of respondents and broadband adoption (Table 6.14).

Regression Analysis I

A regression analysis was performed with behavioural intention as the dependent variable and relative advantage, utilitarian outcomes, hedonic outcomes, primary influence, facilitating conditions resources, knowledge, and self-efficacy as the predictor variables. A total of 358 cases were analysed. From the analysis, a significant model emerged (F (7, 358) = 40.576, $p < 0.001$) with the adjusted R square being 0.437 (Table 6.15). The significant variables are shown in Table 6.16 and include FCR (β = .169, $p < .001$), HO (β = .094, $p = .027$), PI (β = .196, $p < .001$), SE (β = .139, $p = .005$) and RA (β = .230, $p < .001$). Knowledge (β = .086, $p = .121$) and utilitarian outcomes (β = .098, $p = .072$) were not considered to be significant predictors in this model.

Regression Analysis II: After Removing Knowledge Constructs from Predictors

Knowledge and utilitarian outcomes were not significant predictors in the model obtained from the regression analysis whose results are presented in Table 6.16.

Table 6.15. Regression analysis I: Model summary

Model	R	R²	Adjusted R²	Std. Error of the Estimate
1	.669(a)	.448	.437	1.10130

a. Predictors: (Constant), FCR, HO, PI, SE, RA, UO, K

Table 6.16. Regression analysis I: Coefficients [a]

	Unstandardized Coefficients		Standardized Coefficients				Collinearity Statistics	
	B	Std. Error	β	t	p	Partial Correlations	Tolerance	VIF
(Constant)	-.830	.434		-1.913	.057			
RA	.348	.081	.230	4.299	.000	.224	.549	1.820
UO	.132	.073	.098	1.807	.072	.096	.537	1.861
HO	.091	.041	.094	2.216	.027	.118	.880	1.136
PI	.197	.045	.196	4.395	.000	.229	.789	1.267
K	.099	.064	.086	1.553	.121	.083	.511	1.957
SE	.174	.061	.139	2.837	.005	.150	.655	1.527
FCR	.179	.049	.169	3.635	.000	.191	.731	1.368

[a.] Dependent Variable: BI

Table 6.17. Regression analysis II: Model summary

Model	R	R^2	Adjusted R^2	Std. Error of the Estimate
1	.666[a]	.444	.435	1.10351

[a.] Predictors: (Constant), FCR, HO, PI, SE, RA, UO

The p value of the utilitarian outcomes construct was close to the significance level; however, for the knowledge construct it was not. Therefore, it was decided to undertake another regression analysis cycle keeping the other settings as above but removing knowledge from the predictors list.

The regression analysis was performed with behavioural intention as the dependent variable and relative advantage, utilitarian outcomes, hedonic outcomes, primary influence, facilitating conditions resources and self-efficacy as the predictor variables. This time the total number of predictor variables included in the analysis was six, which was one less than before as knowledge was eliminated from the list. A total of 358 cases were analysed. From the analysis, a significant model emerged (F (6, 358) = 46.749, $p < .001$). The adjusted R square was 0.435 (Table 6.17). This time all six including the utilitarian outcomes predictor variables included in the second round of analysis were found to be significant (Table 6.18). These include FCR (β

= .169, p < .001), HO (β = .100, p = .018), PI (β = .195, p < .001), SE (β = .165, p < .001), RA (β = .255, p < .001) and utilitarian outcomes (β = .113, p = .035).

As illustrated in Table 6.18, the constructs are arranged according to their size of β values in decreasing order. The size of β suggests that relative advantage has the largest impact in the explanation of variations of BI. This is followed by the primary influence construct and then facilitating conditions resources. This suggests that the first three constructs that have the largest impact in explaining variance of BI belong to all three categories (i.e., attitudinal, normative and control constructs). The self-efficacy construct from the control category contributed the fourth largest variance of BI. The remaining two constructs (e.g., UO and HO) were from the attitudinal category.

When performing a regression analysis, an important cause for concern is the existence of multicollinearity amongst the independent variables such as RA, PI, FCR, SE, UO, and HO. It is likely to exist when the independent variables included in the analysis are not truly independent and measure redundant information (Myers, 1990).

The existence of multicollinearity negatively affects the predictive ability of the regression model (Myers, 1990) and causes problems when attempting to draw inferences about the relative contribution of each predictor variable to the success of a model (Brace et al., 2003). Therefore, it is important to examine whether the problem of multicollinearity exists in this research.

SPSS provides two options to estimate the tolerance and variance inflation factor (VIF) to trace if data suffers with the problem of multicollinearity (Brace et al., 2003; Myers, 1990). According to Myers (1990), if the VIF value for any constructs

Table 6.18. Regression analysis II: Coefficients [a]

	Unstandardized Coefficients		Standardized Coefficients			Partial Correlations	Collinearity Statistics	
	B	Std. Error	β	t	p		Tolerance	VIF
(Constant)	-.904	.432		-2.093	.037			
RA	.384	.077	.255	4.962	.000	.256	.601	1.664
PI	.196	.045	.195	4.362	.000	.227	.789	1.267
FCR	.191	.049	.180	3.916	.000	.205	.749	1.335
SE	.206	.058	.165	3.582	.000	.188	.742	1.347
UO	.153	.072	.113	2.116	.035	.112	.555	1.801
HO	.097	.041	.100	2.382	.018	.126	.890	1.124

[a] Dependent Variable: BI

surpasses 10, then there is a possibility of multicollinearity amongst constructs. If detected, in order to overcome this problem, a variable with a VIF value more than 10 needs to be deleted (Myers, 1990).

An alternative to this approach is an estimation of the tolerance value. The tolerance values are a measure of the correlation between the predictor variables and vary between 0 and 1. The closer to zero the tolerance value is for a variable, the stronger the relationship between this and the other predictor variables. It is a matter of concern if a predictor variable in a model has a tolerance of less than 0.0001 (Brace et al., 2003).

In order to detect multicollinearity in this research, both the VIF and tolerance that were estimated are shown in Table 6.18. Values obtained for both VIF and tolerance indicate that there is no problem of multicollinearity in this research. Table 6.18 illustrates that the VIF for this model varied between 1.80 for primary influence constructs and 1.12 for hedonic outcomes constructs, which are far below the recommended level (Brace et al., 2003; Myers, 1990; Stevens, 1996).

Table 6.18 also illustrates that all the predictors have a high tolerance of more than 0.55. Therefore, both the VIF and tolerance values suggest that the independent variables (i.e., RA, PI, FCR, SE, UO, and HO) included in this study do not suffer from the problem of multicollinearity.

Regression Analysis III: Examining the Relationship Between Overall Attitudinal, Normative, Control Constructs, and Behavioural Intentions

A new scale (i.e., aggregated measure) was created for each attitudinal, normative, and control category. The computing average of all the items for each category achieved this. The purpose was to conduct a regression analysis with behavioural intention as the dependent variable and attitudinal, normative and control as the predictor variables. A total of 358 cases were analysed. From the analysis, once again a significant model emerged (F (3, 358) = 86.932, $p < 0.001$). The adjusted R square was 0.419 (Table 6.19). All three variables were found to be significant

Table 6.19. Regression analysis III: Model summary

Model	R	R^2	Adjusted R^2	Std. Error of the Estimate
1	.651[a]	.424	.419	1.11839

[a] Predictors: (Constant), O_C_CONS, O_N_CONS, O_A_CONS

Table 6.20. Regression analysis III: Coefficients [a]

Model	Predictors	Un standardized Coefficients		Standardized Coefficients	t	p
		B	Std. Error	β		
1	(Constant)	-.230	.388		-.593	.553
	O_A_CONS	.463	.089	.282	5.225	.000
	O_N_CONS	.175	.053	.151	3.318	.001
	O_C_CONS	.525	.071	.367	7.433	.000

[a] Dependent Variable: BI

(shown in Table 6.20). These include the attitudinal (O_A_CONS) (β = .282, $p <$ .001), normative (O_N_CONS) (β = .151, p = .001), and control (O_C_CONS) (β = .367, $p <$ .001).

Logistic Regression: Examining the Relationship between Behavioural Intention, Facilitating Conditions Resources, and Broadband Adoption Behaviour

The dependent variable, which measures the broadband adoption behaviour, is categorical in nature and represented by Yes and No. Yes is equal to 1 if the respondent possesses broadband and 0 if they do not have broadband. It was also possible to employ the ordinary least squares regression to fit a linear probability model. However, the limitation of the linear probability model is that it may predict probability values beyond the 0.1 range; therefore, the logistic regression model was found most appropriate to estimate the factors which influence broadband adoption behaviour (Greene, 1997; Stynes & Peterson, 1984).

A logistic regression analysis was performed with broadband adoption as the dependent variable and behavioural intention and facilitating conditions resources as the predictor variables. A total of 358 cases were analysed and the full model was considered to be significantly reliable (χ^2 (2, N = 358) = 128.559, $p <$ .001) (Table 6.21). This model accounted for between 30.2 percent and 40.6 percent of the variance in broadband adoption (Table 6.22), and 88.4 percent of the broadband adopters were successfully predicted (Table 6.23). However, only 58.9 percent of the predictions for the non-adopters were accurate. Overall, 76.0 percent of the predictions were accurate (Table 6.23).

Table 6.24 offers the coefficients, Wald statistics, associated degrees of freedom, and probability values for each of the predictor variables. This shows that both the behavioural intention (BI) and facilitating conditions resources (FCR) reliably

Table 6.21. Logistic regression: Omnibus tests of model coefficients

		χ^2	df	p
Step 1	Step	128.559	2	.000
	Block	128.559	2	.000
	Model	128.559	2	.000

Table 6.22. Logistic regression: Model summary

Step	Cox & Snell R^2	Nagelkerke R^2
1	.302	.406

Table 6.23. Logistic regression: Classification table

Observed		Predicted		
		Broadband Adopters and Non Adopters		Percentage Correct
		No	Yes	
Broadband Adopters and Non Adopters	**No**	89	62	58.9
	Yes	24	183	88.4
Overall Percentage				76.0

Table 6.24. Logistic regression: Variables in the equation

		B	S.E.	Wald	df	p	Exp (B)
Step 1[a]	BI	.916	.141	42.021	1	.000	2.500
	FCR	.455	.109	17.471	1	.000	1.576
	Constant	-7.529	.954	62.222	1	.000	.001

[a] Variable(s) entered on step 1: BI, FCR.

predicted broadband adoption. The values of the coefficients reveal that each unit increases in BI and the FCR score is associated with an increase in the odds of broadband adoption by a factor of 2.50 and 1.58 respectively (Table 6.24). This means that BI has a larger part in explaining actual adoption than FCR.

Regression Analysis IV: Explaining the Relatioship Between Service Quality, Secondary Influence and Behavioural Intention to Change Service Provider

A regression analysis was conducted with behavioural intention to change service provider (BISP) as the dependent variable and secondary influence and service quality as predictor variables. A total of 308 cases were analysed. From the analysis, a significant model emerged (F (2, 308) = 13.239, $p < .001$). The adjusted R square was 0.074 (Table 6.25). Both the variables were found to be significant (Table 6.26). These include secondary influence (SI) ($\beta = .153, p = .006$) and service quality (SQ) ($\beta = -.255, p < .001$). Service quality is negatively correlated with the behavioural intention to change service provider, which means that the lower the quality of the service provided, the higher the chance that consumers will change service providers. However, it is important to indicate that since the adjusted R square is very low (Table 6.25), service quality and secondary influence are almost unable to explain the variation of BISP.

Usage of Broadband

Consumers' Online Habits: Rate of Internet Use

Table 6.27 illustrates the difference between broadband and narrowband consumers in terms of the frequency of usage or accessibility to the Internet. The results

Table 6.25. Regression analysis IV: Model summary

Model	R	R²	Adjusted R²	Std. Error of the Estimate
1	.283(a)	.080	.074	1.811

a. Predictors: (Constant), SI, SQ

Table 6.26. Regression analysis IV: Coefficients [a]

Model	Predictors	Un standardized Coefficients		Standardized Coefficients	t	p
		B	Std. Error	β		
1	(Constant)	4.495	.415		10.843	.000
	SQ	-.359	.078	-.255	-4.619	.000
	SI	.166	.060	.153	2.774	.006

[a.] Dependent Variable: BISP

Table 6.27. Frequency of home internet access

Frequency of Internet access	Narrowband		Broadband	
	Frequency	Percent	Frequency	Percent
Several times a day	53	52.5	154	74.4
About once a day	21	20.8	31	15
3-5 days a week	11	10.9	12	5.8
1-2 days a week	10	9.9	8	3.9
Once every few weeks	4	4	2	2
Less often	2	2	0	0
Total	101		207	

χ^2 Test (N=308) Type of connection X Frequency of Internet Access			
	Value	df	p (2-sided)
Pearson χ^2	20.027	5	.001

indicate clear differences and suggest that the majority of broadband consumers (74.4 percent) access or use the Internet several times a day in comparison to 52.5 percent of the narrowband consumers. However, the numbers of broadband consumers decrease as the frequency of Internet access decreases. Only 15 percent of broadband's consumers in comparison to 20.8 percent narrowband consumers access the Internet about once a day. Similarly, 10.9 percent of narrowband consumers access the Internet 3-5 days a week, in comparison to 5.8 percent of broadband users. Generally, broadband consumers' online habits in terms of their frequency of Internet access differ from narrowband consumers. Broadband consumers belong to the more frequent categories whilst narrowband consumers belong to the less frequent categories (Table 6.27). The chi-square test confirmed a significant difference (χ^2 $(5, N = 308) = 20.027, p = .001$) between narrowband and broadband consumers in terms of the frequency of Internet access (Table 6.27).

A binary correlation test was also conducted to examine if there was an association between frequency of Internet access and broadband adoption. The results obtained from this test suggest that there was a significant negative correlation between frequency of Internet access and broadband adoption (Table 6.28).

Table 6.29 illustrates the difference between broadband and narrowband consumers in terms of total time spent on the Internet on a daily basis. Similar to the frequency of Internet access, the results indicate that clear differences occur between narrowband and broadband consumers. Generally, broadband consumers increase as the number of hours increase. Contrastingly, the number of narrowband consumers increase as the hours decrease. Twenty-four percent of narrowband consumers spend less than half an hour in contrast to only 9.2 percent broadband consumers. However, in the 3-4 hours category, broadband consumers (23.2 percent) exceeded the narrowband consumers (17 percent). Sixteen percent of broadband consumers spent more than

Table 6.28. Spearman's rho correlations to show association between duration of internet access and broadband adoption

		Broadband Adoption
Frequency of internet access	Correlation Coefficient	.297(**)
**Correlation is significant at the 0.01 level (1-tailed).	Sig. (1-tailed)	.000
	N	354

Table 6.29. Duration of Internet access on a daily basis

Duration of Internet access	Narrowband		Broadband	
	Frequency	**Percent**	**Frequency**	**Percent**
<1/2 hour	24	24	19	9.2
1/2-1 hour	12	12	15	8.8
>1-2 hour	29	29	68	32.9
>2-3 hour	6	6	22	10.6
>3-4 hour	17	17	48	23.2
=>4 hour	12	12	35	16.9
Total	100		207	
χ^2**Test (N=307)** Type of connection X Duration of Internet Access				
	Value	**df**	**p (2-sided)**	
Pearson χ^2	16.488	5	.006	

four hours on the Internet on a daily basis, in comparison to 12 percent of narrowband users (Table 6.29). The chi-square test confirmed a significant difference (χ^2 (5, $N = 307$) = 16.488, $p = .006$) between narrowband and broadband consumers in terms of the total time spent on the Internet on a daily basis (Table 6.29).

A binary correlation test was also conducted to examine if there was any association between duration of Internet access and broadband adoption. The results obtained from this test suggest that there was a significant negative correlation between duration of Internet access and broadband adoption (Table 6.30).

Variety of Internet Use

The variety of Internet use was computed by counting how many online services the narrowband and broadband consumers use on average. Table 6.31 illustrates that broadband consumers access or use more online services than narrowband consumers. The results indicate that, on average, non-adopters of broadband (i.e., narrowband users) use 17.97 online services, which is significantly lower ($t = 4.107$, df = 273, $p < .001$) than the 22.41 online services used on average by broadband adopters (Table 6.31).

Table 6.30. Spearman's rho correlations to show association between duration of internet access and broadband adoption

		Broadband Adoption
Duration of internet access	Correlation Coefficient	.225(**)
**Correlation is significant at the 0.01 level (1-tailed).	Sig. (1-tailed)	.000
	N	353

Table 6.31. Variety of Internet activities accessed by broadband and narrowband consumers

Type of connection	N	Mean	SD	Std. Error Mean	t	df	p
Narrowband	89	17.97	8.340	.884			
Broadband	186	22.41	8.418	.617	4.107	273	.000

Table 6.32. Spearman's rho correlations to show association between variety of Internet use and type of internet connection

		Type of Internet Connection
Variety of Internet Use	Correlation Coefficient	.256(**)
** Correlation is significant at the 0.01 level (1-tailed).	Sig. (1-tailed)	.000
	N	275

A binary correlation test was also conducted to examine if there was any association between variety of Internet use and the type of Internet connection. The results obtained from this test suggest that there was a significant positive correlation between variety of Internet use and the type of Internet connection (Table 6.32).

Usage of Online Services by Narrowband and Broadband Consumers

A total of 41 online services that belonged to nine different categories (Horrigan & Rainie, 2002) were included to examine the difference in the usage of the Internet by consumers of narrowband and broadband (Table 6.33). These nine categories (Horrigan, & Rainie, 2002) comprised communications (five online services), information seeking (seven online services), information producing (four online services), downloading (six services), media streaming (five services), e-commerce (eight services), entertainment activities (four services), social and personal (two services), and e-government (Table 6.33).

For all 41 online services, except for e-mail, broadband consumers out-numbered the narrowband consumers. However, the differences between the narrowband and broadband consumers were significant for only 19 online services. The results indicate that use of none of the five online services that were placed within the communications category significantly differed between broadband and narrowband consumers (Table 6.33).

Within the information seeking category, of the seven online services, the use of five was found to be significantly different between broadband and narrowband consumers. These included online news ($\chi^2(1, N = 276) = 6.77, p = .009$), job related research ($\chi^2(1, N = 276) = 13.18, p < .001$), research for school or training ($\chi^2(1, N = 276) = 7.36, p = .007$), searches for travel information ($\chi^2(1, N = 276) = 5.26, p = .002$) and the accessing of online lectures ($\chi^2(1, N = 276) = 3.98, p = .046$).

Within the information producing category, of the four online services, the use of three was found to be significantly different between broadband and narrowband

Table 6.33. Access of online services by broadband and narrowband users (N=276)

Category/Online services	Narrowband Freq.	Narrowband percent	Broadband Freq.	Broadband percent	Total percent	χ^2Test χ^2Value	Df	Sig.
Communications								
Email	89	100	186	99.5	99.6	.478	1	.489
Instant messaging	46	51.7	116	62	58.7	2.66	1	.103
Online Chat	34	38.2	81	43.3	41.7	.649	1	.421
Video conferencing	16	18	49	26.2	23.6	2.26	1	.132
Voice over Internet (VoIP)	18	20.2	57	30.5	27.2	3.20	1	.073
Information Seeking								
Online News	58	65.2	149	79.7	75	6.77	1	.009
Job related research	60	67.4	161	86.1	80.1	13.18	1	.000
Look for product info	78	87.6	171	91.4	90.2	.988	1	.320
Research for school or training	54	60.7	143	76.5	71.4	7.36	1	.007
Look for travel information	75	84.3	174	93	90.2	5.26	1	.002
Look for medical information	61	68.5	140	74.9	72.8	1.22	1	.269
Online lectures	17	19.1	57	30.5	26.8	3.98	1	.046
Information Producing								
Share computer files	37	41.6	111	59.4	53.6	7.67	1	.006
Create content (e.g., Web pages)	30	33.7	70	37.4	36.2	.362	1	.547
Store/display/develop photos	49	55.1	135	72.2	66.7	7.96	1	.005
Store files on the Internet	26	29.2	97	51.9	44.6	12.53	1	.000
Downloading								
Download games	23	25.8	59	31.6	29.7	.941	1	.332
Download video	20	22.5	76	40.6	34.8	8.77	1	.003
Download pictures	46	51.7	120	64.2	60.1	3.92	1	.048
Download music	38	42.7	113	60.4	54.7	7.65	1	.006
Download movie	23	25.8	69	36.9	33.3	3.31	1	.069
Download free software	50	56.2	136	72.7	67.4	7.51	1	.006
Media Streaming								
Video streaming	24	27	84	44.9	39.1	8.16	1	.004
Listen to music (streaming/MP3)	35	39.3	107	57.2	51.4	7.72	1	.005
Listen to the radio station	45	50.6	112	59.9	56.9	2.14	1	.143
Watch movies streaming)	18	20.2	69	36.9	31.5	7.76	1	.005

continued on following page

Table 6.33. continued

E-commerce								
Undertake online banking	43	48.3	123	65.8	60.1	7.67	1	.006
Online bill paying	42	47.2	114	61	56.5	4.65	1	.031
Purchase a product	74	83.1	162	86.1	85.5	.591	1	.442
Purchase a travel service	65	73	148	79.1	77.2	1.27	1	.258
Purchase groceries (household goods)	31	34.8	78	41.7	39.5	1.19	1	.278
Online auctions e.g., eBay	38	42.7	109	58.3	53.3	5.88	1	.015
Buy/sell stocks (online share trading)	14	15.7	41	21.9	19.9	1.450	1	.228
Play lottery	6	6.7	25	13.4	11.2	2.65	1	.103
Entertainment activities								
Obtain information on hobby	54	60.7	144	77.0	71.7	7.93	1	.005
Use it for fun e.g., web surfing	57	64	147	78.6	73.9	6.63	1	.010
Play online game	17	19.1	50	26.7	24.3	1.91	1	.167
View or visit adult content websites	14	15.7	30	16	15.9	.004	1	.947
Social and Personal								
Online dating and matrimonial services	8	9	17	9.1	9.1	.001	1	.987
Collaboration with schoolmates	27	30.3	72	38.5	35.9	1.74	1	.186
E-government								
Accessing e-government services	40	44.9	97	51.9	49.6	1.15	1	.282

consumers. These included sharing computer files ($\chi^2(1, N = 276) = 7.67, p = .006$), store/display/develop photos ($\chi^2(1, N = 276) = 7.96, p = .005$), and storing files on the Internet ($\chi^2(1, N = 276) = 12.53, p < .001$) (Table 6.33).

Within the downloading category, of the six online services, the use of three was found to be significantly different between the broadband and narrowband consumers. These included downloading videos ($\chi^2(1, N = 276) = 8.77, p = .003$), downloading music ($\chi^2(1, N = 276) = 7.65, p = .006$), and downloading free software ($\chi^2(1, N = 276) = 7.51, p = .006$) (Table 6.33).

Within the media streaming category, of the four online services, the use of three was found to be significantly different between broadband and narrowband consumers. This included video streaming ($\chi^2(1, N = 276) = 8.16, p = .004$), listening to music ($\chi^2(1, N = 276) = 7.72, p = .005$), and watching movies streaming ($\chi^2(1, N = 276) = 7.76, p = .005$) (Table 6.33).

Within the e-commerce category, of the eight online services, the use of three was found to be significantly different between broadband and narrowband consumers. These included undertaking online banking ($\chi^2(1, N = 276) = 7.67, p = .006$), online bill playing ($\chi^2(1, N = 276) = 4.65, p = .031$), and online auctions ($\chi^2(1, N = 276) = 5.88, p = .015$) (Table 6.33).

Within the entertainment category, of the four online services, the use of two was found to be significantly different between broadband and narrowband consumers. These included obtaining information on hobbies ($\chi^2(1, N = 276) = 7.93, p = .005$), and using it for fun ($\chi^2(1, N = 276) = 6.63, p = .010$) (Table 6.33).

Similar to the communication category, none of the placed services within the remaining two categories (the first being social and personal and the second being e-government) significantly differed between both broadband and narrowband consumers (Table 6.33).

Impact of Broadband

A cross-sectional analysis was applied to both the narrowband and broadband consumers' usage of time upon various activities (Table 6.34). The results found a clear distinction between the narrowband and broadband consumers in terms of time spent on a total of 20 daily life activities examined within this research. The broadband consumers' behaviour in terms of time spent on various activities is dissimilar to dial up consumers. For example, the television-watching behaviour of 41 percent of the broadband consumers decreased in comparison to 36 percent of the dial up consumers. Similarly, reading newspapers/books/magazines was more affected by broadband use. Compared to 18 percent of the dial up users, 28 percent of the broadband consumers read fewer newspapers/books/magazines in comparison.

Other activities where consumers spent less time than before include in-store shopping, working in the office, and commuting in traffic (Table 6.34). Working in the office had decreased for 25 percent of the broadband consumers in comparison to 4 percent of the dial up consumers. 18 percent of the broadband consumers spent less time when commuting in traffic then 4 percent of the narrowband consumers.

There is a minor distinction between the narrowband and broadband consumers in terms of time allocation pattern for activities such as spending time with family and friends, time spent alone, receiving or making phone calls, and outdoor entertainment (Table 6.34).

The only activity where the time spent had increased was working at home. The working at home behaviour of 53 percent of the broadband consumers increased in comparison to 28 percent of the narrowband consumers (Table 6.34).

As illustrated previously, differences existed between the broadband and narrowband consumers in terms of their time allocation pattern for all the 20 activities examined in this research. However, the chi-square test confirmed that narrowband consumers' time allocation patterns differed significantly to the broadband consumers in only five activities (Table 6.34). These included shopping in-store ($\chi^2(2, N = 278) = 6.7$, $p = .034$), working at home ($\chi^2 (2, N = 278) = 18.1, p < .001$), reading newspapers/books/magazines ($\chi^2(2, N = 278) = 7.6, p = .022$), working in the office ($\chi^2(2, N = 278) = 18.4, p < .001$), and commuting in traffic ($\chi^2(2, N = 278) = 10.8, p = .005$). It was also found that the time allocation patterns for the remaining 15 activities did

Table 6.34. The impact of broadband internet on various daily life activities (N=278)

| Daily Life Activities | Type of Internet Connection | | | | | | χ^2 Tests | | |
| | Narrowband n=92 | | | Broadband n=186 | | | | | |
	Nc. (%)	Dec. (%)	Inc. (%)	Nc. (%)	Dec. (%)	Inc. (%)	χ^2	df	Sig.
Watching television/cable/satellite	62	36	2	56	41	3	.92	2	.629
Shopping in stores	77	16	7	63	31	6	6.7	2	**.034**
Working at home	68	5	28	42	5	53	18.1	2	**.000**
Reading e.g., newspapers/books	76	18	5	60	28	12	7.6	2	**.022**
Working in the office	86	4	10	65	25	10	18.4	2	**.000**
Commuting in traffic	95	4	1	80	18	2	10.8	2	**.005**
Spending time with family	77	17	5	81	12	6	1.33	2	.513
Spending time with friends	80	13	7	82	12	6	.137	2	.934
Attending social events	85	8	8	84	10	6	.567	2	.753
Time spent on sport	82	12	7	83	12	4	.636	2	.728
Time spent on hobbies	77	14	9	73	16	11	.731	2	.694
Time spent on sleeping	72	24	4	77	20	3	1.08	2	.581
Time spent alone (doing nothing)	61	25	14	66	25	9	1.69	2	.430
Studying	60	14	26	63	10	26	.954	2	.621
Household work	82	14	4	82	17	1	3.31	2	.191
Receiving/ making phone calls	63	34	3	60	32	9	2.75	2	.258
Doing charity and social works	86	8	7	89	7	4	.692	2	.707
Outdoor recreation (DIY, pet care)	85	8	8	89	8	3	2.66	2	.263
Outdoor entertainment	85	11	4	76	12	12	4.28	2	.117
Visiting or meeting friends or relatives	85	11	4	85	11	4	.063	2	.969

Legend: NI= Not Included, **Inc.**=Increased, **Dec.**= Decreased, **Nc.**=No Change not differ significantly between the narrowband and broadband users. However, the Internet on its own has begun to influence the daily routine of consumers. This is evident from the findings in Table 6.34, where it was learnt that both the broadband and narrowband consumers influence the time allocation patterns for undertaking daily life activities.

Summary

This chapter presented the findings obtained from the data analysis of the survey that was conducted to examine consumer adoption, usage, and impact of broadband in UK households. The findings were presented in several sections. Findings from the descriptive statistics suggested that all the constructs except the BISP rated strongly (mean above 3.5 at the 1-7 Likert scale). This suggests that the respondents showed strong agreement in factors included in the study for examining the adoption of broadband. There was then an examination of the differences between the adopters and non-adopters of broadband, employing the t-test and discriminant analysis techniques. The results from the t-test and discriminant analysis suggest that significant differences occur between the responses obtained from the narrowband and broadband consumers with regards to attitudinal, normative and control constructs.

Examination of the demographic differences employing the chi-square test suggest that broadband consumers differ significantly to narrowband consumers in terms of age, education, occupation, and income. Finally, the linear and logistic regression analysis provided evidence that the attitudinal, normative and control constructs (independent variable) significantly explain behavioural intentions which, along with the facilitating conditions resources, significantly explain broadband adoption behaviour.

The findings related to the usage of the Internet suggest that broadband consumers significantly differ to narrowband users in terms of their online habits and variety of Internet use. When accessing or using 19 online services from an overall total of 41 services, the numbers of broadband consumers were significantly higher than the narrowband consumers.

The last section of this chapter examined the effects of broadband usage on consumers' time allocation pattern on twenty daily life activities. The findings suggest that for all twenty activities, broadband consumers' time allocation pattern differ to that of narrowband consumers; however, the differences were found to be significant only for five activities.

Chapter 8 will discuss the findings in light of the previous work. It will provide a discussion on the validation of the instrument, model refinement, and the usage and impact of broadband within the context of UK households.

References

Brace, N., Kemp, R., & Snelgar, R. (2003). *SPSS for psychologists: a guide to data analysis using SPSS for windows.* New York: Palgrave Macmillan.

Gilligan, C., & Wilson, R. M. S. (2003). *Strategic marketing planning.* Oxford: Butterworth-Heinemann.

Greene, W. H. (1997). *Econometric analysis.* Prentice Hall.

Myers, R. H. (1990). *Classical and modern regression with applications.* Boston: PWS-KENT Publishing Company.

Rice, C. (1997). *Understanding customers.* Oxford: Butter worth-Heinemann.

Stevens, J. (1996). *Applied multivariate statistics for the social sciences.* NJ: Lawrence Erlbaum Associates, Inc.

Stynes, D. J., & George L. P. (1984). A review of logit models with implications for modeling recreation choices. *Journal of Leisure Research, 16,* 295-310.

Chapter 7

Comparing the Current and Future Use of Electronic Services

Abstract

The previous chapter (Chapter 6) examined the differential usage of the Internet in broadband and narrowband environments. However, the previous chapter excluded the comparison of current and future consumer use of various electronic services and applications at home and in the work place in the UK. A recently published report that was submitted to the Department of Trade and Industry (DTI), UK, suggests that the penetration of broadband is likely to promote the usage of advanced Internet content and applications; however, due to the lack of data at present, it is difficult to support this theoretical claim (Analysys, 2005). The Analysys report states, "Much has been made of the requirement for countries to invest in broadband communications infrastructure, and to promote its usage. Increased take-up of broadband access services is expected to stimulate usage of advanced Internet content and applications by consumers and by businesses, thus changing individuals' behaviour, creating new industries, or increasing productivity in existing industries. However, data to prove the theory is hard to come by" (Analysys, 2005).

This implies that examining the current trend of consumer behaviour towards the uses of various emerging electronic services will not only help to encourage their further adoption and use, but also to promote the adoption of broadband. Utilising this reasoning as a motivating factor, this chapter progresses a step further towards understanding the trend of current and future uses of various online/electronic services at home and in the work place in the UK. Having introduced the aim of the chapter, the next section presents the findings. Finally, a concluding discussion to the research presented in this chapter is also provided. It is important to mention that the research methodology and the theoretical basis for this chapter is already presented and discussed in previous chapters. Therefore, this chapter does not include these elements and, instead, solely presents empirical data in the form of charts.

Adoption and Use of Electronic Services and Applications

A total of 41 online services belonging to seven different categories (see Chapter 6) were included to examine the current and future use of the Internet at home and in the work place in the UK (Figures 7.1-7.7). These seven categories comprised communications (five online services), information seeking (seven online services), information producing (four online services), downloading (six services), media streaming (five services), e-commerce (eight services), and other activities that included entertainment activities (four services), social and personal (two services), and e-government (Figures 7.1-7.7).

Communications

Figure 7.1 indicates that for communication purposes within the home, e-mail was used the most (100 percent), followed by in the workplace (94 percent). The communication related Internet activities that were utilised as categories included instant messaging, online chat, video conferencing and voice over Internet (VoIP). When determining the reported future use of email, it was found that it was slightly lower (99 percent) than the current rate of use at home.

This is because some of the respondents did not provide their responses for the future use of e-mail. Within the communications category, utilising the Internet for instant messaging purposes was the second most widely used online activity both at home (59 percent) and the work place (33 percent). Its future use is also reported to increase as 70 percent of the respondents agreed that they intended to use the Internet for sending instant messages in the future. Comparatively, video conferenc-

Figure 7.1. Current and future use of electronic services within communication category

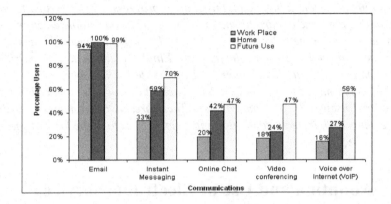

ing and VoIP were still the least widely used communication related activities over the Internet (Figure 7.1). The results of this research also indicated that although an increasing number of consumers intended to use all five of the e-services in future, VoIP was likely to experience a larger rate of adoption than the other four activities (Figure 7.1).

Information Seeking Activities

Figure 7.2 illustrates that for information seeking purposes, the search for products and travel information services were used mostly by the majority of home dwelling respondents (90 percent). The reported future use of product search is slightly lower (88 percent) than the current use at home. This is because some of the respondents did not provide their responses for future use. Within this category, utilising the Internet for job related research was the second most widely used online activity in the home (80 percent) and slightly lower at work (71 percent). Its future use is also reported to increase as 85 percent of the respondents agreed that they intend to use the Internet for this purpose in the future. Accessing online lectures and researching for medical information were the least widely used information seeking related activities over the Internet (Figure 7.2). The findings of this research also indicated that although more consumers intended to use all seven of the e-services in future, online lectures are likely to experience larger adoption rates than the other four activities (Figure 7.2).

Figure 7.2. Current and future use of electronic services within information seeking category

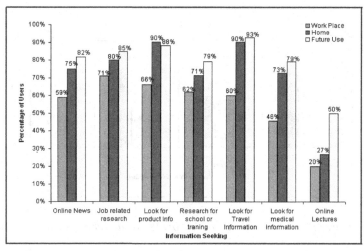

Information Producing Activities

Figure 7.3 shows that within the information producing category, the store/display/develop photos functions were the foremost used activities at home (67 percent) with a much lesser amount in the workplace (33 percent). Its future use is also reported to increase as 76 percent of the respondents agreed that they intended to use the Internet for storing, displaying and developing photos. Comparatively, creating content, for example Web pages, was still the least widely used activity as only 36 percent of the respondents performed this activity at home and an even lesser amount in the work place (27 percent) (Figure 7.3). The findings also indicated that although more consumers intended to use all four of the e-services in future, creating content such as web pages is likely to experience a larger adoption rate than the other three activities (Figure 7.3).

Downloading Activities

Figure 7.4 shows that for downloading purposes, downloading free software was used the most at home (67 percent) and also in the workplace (37 percent). Its future use also reported an increase as 75 percent of the respondents agreed that they intended to use the Internet for downloading free software in future. Within this category,

Figure 7.3. Current and future use of electronic services within information producing category

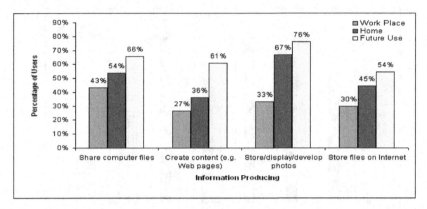

Figure 7.4. Current and future trend for use of electronic services within downloading category

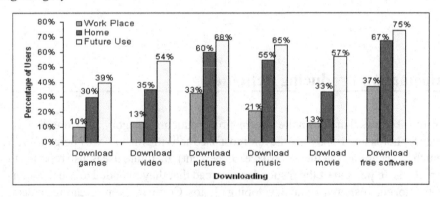

utilising the Internet for downloading pictures was the second most widely used online activity both at home (60 percent) and in the work place (33 percent). Its future use is also reported to increase as 68 percent of the respondents agreed that they intended to use the Internet for this purpose in the future. Additionally, within this category, downloading games was still the least widely used activity (Figure 7.4). The results also indicated that although an increasing number of consumers intended to use all six e-services in the future, downloading movies and videos were likely to experience larger adoption rates than the other four activities (Figure 7.4).

Figure 7.5. Current and future use of electronic services within media streaming category

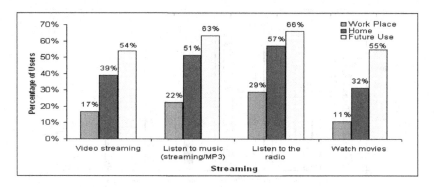

Media Streaming Activities

In Figure 7.5, it can be learnt that for media streaming purposes, the radio was listened to the most at home (57 percent) and in the workplace (29 percent). Its future use is also reported to increase as 66 percent of the respondents agreed that they intended to use the Internet for listening to the radio in future. Within this category, utilising the Internet for listening to music (audio streaming) was the second most widely used online activity both at home (51 percent) and in the work place (22 percent). Its future use is also reported to increase as 63 percent of the respondents agreed that they intended to use the Internet for this purpose in future. Within this category, watching movies and video streaming were still the least widely used activities (Figure 7.5). The findings also indicate that although an increasing number of consumers intended to use all four e-services in the future, watching movies was likely to experience larger adoption rates than the other three activities (Figure 7.5).

E-Commerce

Figure 7.6 illustrates that for e-commerce purposes, the purchasing of a product was higher within the home (86 percent) than in the workplace (46 percent). The reported future use of product purchase is slightly lower (85 percent) in the context of homes. This is because some of the respondents did not provide their responses for the future use of product purchases. Within this category, utilising the Internet for purchasing travel tickets was the second most widely used online activity both at home (77 percent) and in the work place (44 percent). Its future use is also reported to increase as 82 percent of the respondents agreed that they intended to use

Figure 7.6. Current and future use of electronic services within e-commerce category

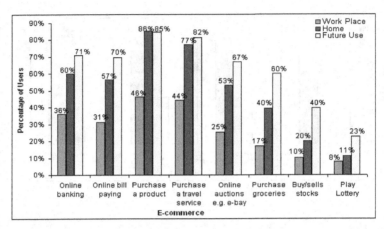

the Internet for future purposes. In comparison, activities such as the buying and selling of stocks and shares (20 percent) and lottery playing were the least widely used e-commerce related activities over the Internet (Figure 7.6). The findings also indicated that although large numbers of consumers intended to use all of the eight e-services in the future, the purchase and selling of stocks and shares and groceries purchasing were likely to experience larger adoption rates than the other six activities (Figure 7.6).

Other Online Activities

Figure 7.7 illustrates that within this category, the use of the Internet for fun, for example Web surfing, was foremost used at home (74 percent) and in the workplace (46 percent). Its future use is also reported to increase as 77 percent of the respondents agreed that they intended to use the Internet for sending instant messages in the future. Within this category, utilising the Internet for the purpose of obtaining information on hobbies was the highest both in the home (72 percent) and in the work place (39 percent). Its future use is also reported to increase as 78 percent of the respondents agreed that they intended to use the Internet for sending instant messages in the future. Online dating and matrimonial services were the least widely used activities within this category (Figure 7.7). The results of this research also indicated that although more consumers intended to use all seven e-services in the future, accessing e-government services and playing online games were likely to experience larger adoption rates than the other five activities (Figure 7.7).

Figure 7.7. Current and future use of electronic services within other online activities category

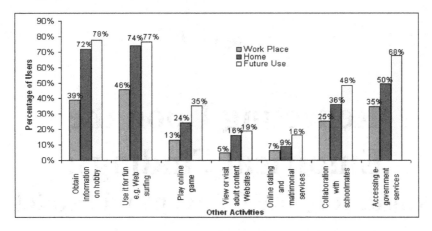

Conclusion

The adoption and use of 41 electronic services and applications in the UK were presented within this chapter. The chapter concluded that the email, search, storing and displaying photos, downloading free software, listening to the radio, product purchasing, and using the Internet for fun emerged as the most commonly used online activities. Instant messaging, job-related researches, sharing computer files, online news, downloading online music, purchasing travel services, online banking, and accessing e-government services are substantially used activities in an online environment. In contrast, video conferencing, video streaming, on-line games, and video downloading are the least adopted online services amongst the UK's Internet users. The chapter concluded that traditional online activities such as email and search have become part of the daily life, while other activities such as online banking, share-market trading, education, and online music have become substantially popular. Alternatively, more technically advanced services such as video conferencing and streaming are in the initial stages of popularity amongst Internet users.

References

Analysys Report (2005). Sophisticated broadband services. *Final Report for the Department of Trade and Industry.* Retrieved July 15, 2005, from http://www. egovmonitor.com/reports/rep11610.pdf

Chapter 8

Reflecting Upon the Empirical Findings:
Validating the Conceptual Model

Abstract

The previous chapters (Chapters 6 and 7) presented the findings obtained from the survey conducted to examine the adoption, usage, and impact of broadband in UK households. The purpose of this chapter is to discuss and reflect upon the findings from a theoretical perspective using those presented in Chapter 2. It also discusses the empirical issues that have been reported from the survey findings in the previous chapter. This chapter is structured as follows. A summary of the hypotheses test is provided and discussed in the next section. This is followed by a discussion and reflection upon the conceptual model of broadband adoption developed within this research. The usage of broadband and its effects on consumers' time allocation patterns on various daily life activities are then discussed and illustrated. Finally, the summary and conclusions of the chapter are provided in the ultimate section.

Research Hypotheses

Although the explanation and discussion on each hypothesis included in this study are provided in the following sections, this section summarises the numbers of hypotheses proposed in Chapter 2 and states whether they are supported by the data or not. Table 8.1 illustrates that a total of 14 research hypotheses were tested to examine whether the independent variables significantly explained the dependent variables. Of the 14 research hypotheses, only one (*H11*) was not supported by the data. The fact that the remaining 13 research hypotheses were supported by the data means that all but one independent variable significantly explained consumers' intention to adopt broadband. Further discussions on the 14 research hypotheses are provided in the following sections. In order to examine the demographic differences between broadband and narrowband consumers, a total of five research hypotheses were tested. These five research hypotheses relating to the differences between the broadband and narrowband consumers were supported by the data (Table 8.1) and further discussions are provided in the following section.

To examine the usage related differences between broadband adopters and non-adopters, three research hypotheses (*H19a, H19b, H19c*) were tested and all the data supported each of the three hypotheses (Table 8.1). Further discussions on the three usage-related research hypotheses are provided in the following sections.

Broadband Adoption

Attitudinal Constructs

As discussed in Chapter 2, if the attitude of individuals towards technology adoption behaviour is positive, then they are likely to form an intention to perform the behaviour (Ajzen, 1985, 1991; Fishbein & Ajzen, 1975; Tan & Teo, 2000; Taylor & Todd, 1995). Following this idea, it was assumed that if the perception of the respondents regarding the attitudinal factor is positive, then it is more likely that it will have a positive influence on their behavioural intention. This theoretical assumption is confirmed by the findings obtained in this research, which suggest that the overall attitudinal factors have a significant positive influence on the behavioural intention to adopt broadband (Figure 8.1).

Following the theoretical basis presented in Chapter 2 (Taylor & Todd, 1995; Venkatesh & Brown, 2001), this research decomposed attitude into four dimensions: hedonic outcomes, utilitarian outcomes (Venkatesh & Brown, 2001), relative advantage (Rogers, 1995), and service quality. Three constructs, namely relative

Table 8.1. Summary of research hypotheses

HN	Research Hypotheses	Results
H1	Overall attitudinal factors will have a positive influence on the behavioural intention to adopt broadband	Supported
H2	Relative advantage will have a positive influence on behavioural intention	Supported
H3	Utilitarian outcomes will have a positive influence on behavioural intention.	Supported
H4	Hedonic outcomes will have a positive influence on behavioural intention.	Supported
H5	Service quality will have a negative influence on the behavioural intention when changing from a current service provider.	Supported
H6	Overall the normative factors will have a positive influence on the behavioural intention when adopting broadband.	Supported
H7	Primary influences will have a positive influence on the perceived behavioural intention to adopt broadband.	Supported
H8	Secondary influences will have positive influence on perceived behavioural intention to change current service providers.	Supported
H9	The overall control factors will have a positive influence on the behavioural intention to adopt broadband.	Supported
H10a	Facilitating conditions resources will have a positive influence on the behavioural intention to adopt broadband.	Supported
H10b	Facilitating conditions resources will have a positive influence on the adoption of broadband.	Supported
H11	Knowledge will have a positive influence on the behavioural intention to adopt broadband.	**Not Supported**
H12	Self-efficacy will have a positive influence on the behavioural intention to adopt broadband.	Supported
H13	Behavioural intention and facilitating conditions resources will have an influence on the adoption of broadband.	Supported
H14	There will be a difference between the adopters and non-adopters of the various age groups.	Supported
H15	The adopters of broadband will be more from male than female gender.	**Not Supported**
H16	There will be a difference between the adopters and non-adopters of broadband in different levels of education.	Supported
H17	There will be a difference between the adopters and non-adopters of different levels of household annual income.	Supported
H18	There will be a difference between the adopters and non-adopters of different types of occupation.	Supported
H19a	The adopters of broadband will spend more time online than non-adopters.	Supported
H19b	The adopters of broadband will access the Internet more frequently than non-adopters	Supported
H19c	The adopters of broadband will access a higher number of online activities than the non-adopters.	Supported

advantage, utilitarian outcomes, and hedonic outcomes, were expected to provide measures of attitude towards the behaviour of broadband adoption in UK households. The fourth construct, service quality, was expected to predict if the adopters are contracted or obligated to the same broadband provider. Alternatively, if the adopters were not satisfied with the obtained service, they will switch to another provider. The outcomes obtained from this study regarding the attitudinal factors are discussed in detail later.

Relative Advantage

As discussed in Chapter 2, several previous empirical studies have found that perceived relative advantage is an important factor for determining the adoption of an innovation (Tan & Teo, 2000; Taylor & Todd, 1995; Tornatzky & Klein, 1982). In comparison to narrowband, broadband offers faster, un-metered, always-on access to the Internet, and provides a number of advantages, convenience, and satisfaction to its users. It was expected that individuals who perceive broadband as advantageous would also be likely to adopt the technology. The findings obtained in this study confirmed that relative advantage has a significant positive influence on the behavioural intention to adopt broadband (Tables 6.18 and 8.1). This study also confirmed that the non-adopters' (i.e., narrowband consumers) score for perceived relative advantage of having broadband is significantly lower than the adopters of broadband (Table 6.4). This is in line with the diffusion theory and previous work on technology adoption and diffusion (Moore & Benbasat, 1991).

Utilitarian Outcomes

As defined in Chapter 2, utilitarian outcomes referred to the extent to which using a technology enhances the effectiveness of household activities such as budgeting, homework, and office work (Venkatesh & Brown, 2001). This construct was proposed and validated to examine the adoption of technology (i.e., PC) in a household setting (Venkatesh & Brown, 2001). Theoretically, it has been argued that broadband can offer a more flexible lifestyle (BSG, 2004). For instance, many people subscribe to broadband in order to work at home instead of travelling to the office; broadband can assist the children with their homework, and many more household activities can be performed conveniently using the faster access of the Internet offered via broadband. Therefore, it is expected that the greater the perception of the usefulness of broadband for work or household related activities, the more likely it is that broadband technology will be adopted in the home. The findings obtained from this study are consistent with this assumption; the findings confirmed that the perceived utilitarian outcomes construct has a significant positive influence on the behavioural intention to adopt broadband (Tables 6.18 and 8.1). It was also found that non-adopters scored significantly lower than adopters on perceived utilitarian outcomes (Table 6.4).

Hedonic Outcomes

Venkatesh and Brown (2001) defined hedonic outcomes as pleasure derived from PC use; for example, for games, fun, and entertainment. Hedonic information systems are described as a self-fulfilling activity and strongly connected to the home and leisure activities, focused on the fun aspect of using information systems, encouraging prolonged rather than productive use (Heijden, 2004). Empirical findings from the Venkatesh and Brown (2001) study established that, when adopting a technology, the role of entertainment (PC games and video games) was important as a factor for consideration in the consumer decision-making process (Venkatesh & Brown, 2001). The entertainment potential of a PC was much more enhanced by the advent of the Internet (Venkatesh & Brown, 2001). The Internet offered the opportunity to play online games, download music and video, chat, and send online messages (Venkatesh & Brown, 2001). However, this potential was severely hampered by the slow speed of dial up Internet (Rose et al., 1999). This barrier is being overcome by broadband technology, which offers faster download speeds and streaming capabilities to Internet users, and hence more convenience and compelling environments (Anderson et al., 2002; BSG, 2004). Considering the entertainment potential that broadband offers in comparison to narrowband, it was expected that individuals who perceive broadband as a good entertainment medium will also be likely to adopt the technology. The findings confirm the underlying research hypotheses that states hedonic outcomes will have a significant positive influence on the behavioural intention (Tables 6.18 and 8.1). This is in line with the findings of recent studies (Lee & Choudrie, 2002; Lee et al., 2003) that suggested that an important factor that was responsible for broadband adoption in South Korea was the PC bang phenomenon. It also supported Anderson et al. (2002), which argued that broadband users are more likely to use the Internet for fun and entertainment in comparison to narrowband users.

The findings also suggested that the non-adopters scored significantly lower than the adopters on perceived hedonic outcomes (Table 6.4). However, an important issue that was observed from the findings was that the agreement of respondents on perceived hedonic outcomes of both broadband and narrowband consumers were much lower than the two attitudinal constructs- relative advantage and utilitarian outcomes (Table 6.3). A possible reason for this is the recent legal restriction against freeloading of music from the Internet (Anderson, 2000; Bhattacharjee et al., 2003; Cowen, 2004; Premkumar, 2003). The restriction is likely to reduce the impact on consumer attitude towards using broadband for entertainment purposes. Freeloading and peer-to-peer online sharing of music is considered similar to software piracy

(Bhattacharjee et al., 2003) and considered as life threatening to the music industry (Premkumar, 2003).

Since the early stages of high speed Internet diffusion, digital freeloading of music has become a cause for concern for regulators and law makers in the UK and the rest of Europe (Anderson, 2000). The sensitivity of the piracy issue became evident when the music industry across Europe, including the UK, sued hundreds of consumers who were engaged in sharing music files on the Internet (Cowen, 2004). A study that examined the relationship between regulations, information technologies, and human behaviour found that regulation does affect the human behaviour of file sharing in peer-to-peer applications (Mlcakova & Whiteley, 2004). Therefore, such legal regulation may hinder consumers realising the entertainment potential of broadband. This was considered to be a plausible reason for why respondents of this study considered hedonic outcomes as less important.

Service Quality

It was discussed in Chapter 2 that only a limited number of studies have included the service quality construct to measure the successful adoption of technology. DeLone and McLean (2003) extended the IS success model (DeLone & McLean, 1992) by integrating a service quality construct. This construct was included to evaluate the fact that an IS department also has a role in facilitating end-user computing via the services that are offered to business personnel wishing to develop their own systems (Rosemann & Vessey, 2005). However, this construct was not employed in the case of PC adoption study in the household context (Venkatesh & Brown, 2001). This is because when purchasing PC, consumers have only one opportunity to make a choice: to purchase or not to purchase. And once a product is sold, the seller is not expected to provide any further after-sales customer support. However, the case of a broadband subscription is different to a PC purchase. That is, the consumers sign an annual contract and during this period, if the provided service is not satisfactory, they can discontinue the broadband subscription.

Alternatively, if consumers have a choice of providers then they might transfer to the competitors. Therefore, it is important to understand whether consumers are satisfied with their current providers and provided services. The findings of this study suggest that service quality has a significant negative influence on the behavioural intention to change the current service provider (Tables 6.18 and 8.1). This means that if the consumer perception is lower for the quality of service obtained from the current service provider, s/he is more likely to switch to new providers. It was also expected that the non-adopters would score significantly lower than the adopters of broadband on service quality, which is confirmed by the findings obtained from t-test (Table 6.4).

Normative Constructs

As per the discussion provided in Chapter 2, the subjective norm in its original form in the theory of planned behaviour (TPB) is employed as a single dimensional construct and is considered directly related to the behavioural intention. This is because a person's behaviour is based on their perception of what others think of what they should do (Tan & Teo, 2000). Following the theoretical arguments in the existing studies, it was expected that the stronger the perceived social influence to adopt broadband, the more likely it is that consumers will develop a stronger intent to subscribe to broadband (Ajzen, 1985, 1991; Fishbein & Ajzen, 1975, Tan & Teo, 2000; Taylor & Todd, 1995; Venkatesh & Brown, 2001). Findings of this study supported the research hypothesis which confirmed that, overall, the normative factors have a positive influence on the behavioural intention to adopt broadband (Figure 6.2b).

In terms of consumer-oriented service, the sources of influence could be the adopter's friends, family and colleagues/peers (Tan & Teo, 2000). Rice et al. (1990) defined such influence as social pressures where members of a social network affect one another's behaviour. Venkatesh and Brown's (2001) study suggests that social influences are significant determinants of the purchasing behaviour of PCs. Similarly, it was also expected that households with broadband connections are likely to influence their relatives and friends by telling them and demonstrating to them the benefits and convenience offered by broadband. Measures that influence adopters can appear in two forms that are termed as primary and secondary influences (Taylor & Todd, 1995; Venkatesh & Brown, 2001). These two dimensions are separated and defined. The findings obtained from this study on the contributions of the primary and secondary influence factors in explaining behavioural intention to adopt broadband are also discussed in detail.

Primary Influences

Social influence from friends, colleagues, peers, and family members that take the form of conversations and messages, and assist in forming perceptions of broadband adoption, is defined as a primary influence (Venkatesh & Brown, 2001). Considering the findings from the previous studies (Taylor & Todd, 1995; Venkatesh & Brown, 2001), it was expected that if broadband adopters are influenced by their social networks with positive messages, they are more likely to have a strong behavioural intention to adopt broadband. The findings of this study confirmed that primary influences have a statistically significant positive influence on the perceived behavioural intention to adopt broadband (Tables 6.18 and 8.1). It was also examined whether the perception of primary influence differs between narrowband and broadband consumers when regarding the influence to adopt broadband. The findings confirmed that non-adopters differed significantly from adopters in terms of primary influences (Table 6.4).

Secondary Influences

As discussed in Chapter 2, messages that are disseminated using mass media, such as TV and newspaper advertisements (secondary sources of information), are considered to be secondary influences, which are likely to influence consumer's intentions to adopt or reject the technology in question (Rogers, 1995; Venkatesh & Brown, 2001). In terms of this research, a secondary influence affects those who have already adopted broadband but are not satisfied with the service quality. Hence, if there are advertisements on TV or in newspapers which promote broadband packages that are economical and offer a better quality service, they are more likely to cause the adopters to contract with a new provider. This theoretical argument was supported by the findings obtained in this research, since the results illustrate that secondary influences have a positive influence on the perceived behavioural intention to change current service providers (Tables 6.18 and 8.1). However, it was found that non-adopters did not score significantly lower than the adopters in terms of secondary influence (Table 6.4).

Control Constructs

Findings from this study provide evidence that the overall control factors have a significant positive influence on the behavioural intention to adopt broadband (Figure 8.1). The finding is consistent with the TPB, which suggests that the presence of constraints can inhibit both the behavioural intention to perform behaviour and the actual behaviour itself (Ajzen, 1985, 1991). The findings of this research and the aforementioned theoretical argument are in line with the findings of this research that illustrate the higher the perception of an individual's control over their internal and external constraint, the more likely that he/she will adopt the technology in question (Ajzen, 1991; Tan & Teo, 2000). However, if the individual's control over the external and internal constraints is low, then despite having a strong behavioural intention, he/she is less likely to adopt the technology (Ajzen, 1991,1985).

In order to develop a better understanding, consistent with the DTPB and MATH, the current study considered the following three constructs as barriers to the adoption of broadband: high costs (i.e., facilitating conditions resources), the ease/difficulty of PCs, Internet use (i.e., self-efficacy), and the lack of knowledge on broadband's benefits (Mathieson, 1991; Taylor & Todd, 1995; Venkatesh & Brown, 2001). The empirical evidence from this research for the role of these three control constructs for explaining the behavioural intention and actual adoption of broadband is provided.

Facilitating Conditions Resources

The South Korean government's vision recognised an affordable monthly cost of broadband for a middle-income household as an important factor for encouraging high rates of adoption (Lee & Choudrie, 2002). An exploratory study on broadband adoption in the UK also suggests that a high monthly cost is a major barrier that is inhibiting the adoption of broadband in households (Dwivedi et al., 2003). Therefore, it is expected that if the monthly cost to subscribe broadband is perceived as high, then adoption will be slow. Furthermore, broadband technology is not compatible to the specifications of old PCs and necessitates either an upgrade or the purchase of a new PC. However, PCs are not easily replaceable devices for the medium and lower income households.

Therefore an economic barrier in the form of costs that are incurred when upgrading or purchasing new personal computers inhibits the adoption of broadband in the household. In line with the theoretical basis, these findings suggest that the facilitating conditions resources have a significant positive influence on both the behavioural intentions to adopt and the actual adoption of broadband (Tables 6.18 and 8.1). The findings also suggest that the non-adopters scored significantly lower than the adopters on the perceived resources to subscribe to broadband (Table 6.4).

Self-Efficacy

The findings of this study provided evidence that self-efficacy has a positive influence on the behavioural intention to adopt broadband (Tables 6.18 and 8.1). This is because the use of broadband also requires using a PC and the Internet. The ease or difficulty of use and requisite knowledge of a PC and Internet use were expected to have an impact upon broadband adoption. An approach to remove such a barrier can be seen in a country such as South Korea where broadband was successfully deployed. The South Korean government installed a variety of promotion policies, such as "The Ten Million Program," which was designed to boost Internet use amongst housewives, the elderly, military personnel, farmers, as well as excluded social sectors such as low-income families, the disabled and even prisoners (Choudrie & Lee, 2004; Lee et al., 2003; Lee & Choudrie, 2002).

This promotion of providing PC and Internet skills in the year 2000 contributed towards the adoption of the Internet. A total of 4.1 million people, including one million housewives, obtained such skills (Choudrie & Lee, 2004; Lee et al., 2003; Lee & Choudrie, 2002). This initiative led to the removal of barriers of self-efficacy amongst household consumers, which then manifested in large-scale broadband adoption in households within a very short period of time. Although basic skills required for accessing the Internet are similar for both narrowband and broadband consumers, the latter is expected to possess a higher self-efficacy than the earlier one (Oh et al., 2003):

There is little difference between the skills needed for using a modem connection over plain old telephone service networks and broadband connection to the internet. However, when an individual uses the broadband internet, he or she can have access to additional information and services such as video on demand, high-quality MP3 digital music, broadcasting services and other multi-media services. An individual may also need different skills and experience in order to use or to take advantage of the broadband internet connection. Time is also required for a user to learn how to set up the system and what he or she can access via the broadband. (Oh et al., 2003)

This statement clearly suggests that broadband consumers would have a stronger perception of skills in comparison to narrowband consumers. This study provides empirical evidence that the non-adopters (i.e., narrowband consumers) scored significantly lower than the broadband adopters on the perceived skills (Table 6.4).

Knowledge

Rogers (1995) suggested that the level of knowledge about an innovation, its risks and benefits affect its adoption rate. The greater the awareness of the benefits of the innovation amongst the consumers and users, the more likely it is that the innovation gets adopted. Previous research suggests that in South Korea, the consumers were aware of the potentials of broadband (Choudrie & Lee, 2004; Lee & Choudrie, 2002; Lee et al., 2003). The consumers were also aware of the benefits of faster Internet access, which was essential to satisfy their needs. This was considered to be one of the factors that accelerated broadband adoption in South Korea. Therefore, in Chapter 2, it was explained that the adoption of broadband requires a clear message of its usages and benefits amongst the total segments of society (Choudrie & Lee, 2004; Lee & Choudrie, 2002; Lee et al., 2003; Rogers, 1995).

Also, if consumers are not aware of the benefits of adopting a particular innovation, then it is expected that they are more likely to reject the decision to make a purchase due to a lack of the perceived needs. However, in contrast to the theoretical reasoning, the empirical findings suggest that the knowledge construct is unlikely to have a large impact on variance of behavioural intention to adopt broadband (Table 6.16). This is in line with the argument that the majority of consumers are already aware of what to do with the Internet as it permeates people's lives and work environments (Oh et al., 2003). This may be a possible reason why this construct has not contributed largely towards explaining the variance in behavioural intention of adopting broadband.

Research Model of Broadband Adoption (MBA)

A summary of the research hypotheses test, and also a reflection on the hypotheses in relation to the proposed conceptual model, were provided previously. This subsection summarises the discussion and reflects upon the performance of the broadband adoption model in comparison to its guiding models and framework.

The findings suggest that the paths from relative advantage, utilitarian outcomes, and hedonic outcomes towards the behavioural intention to adopt broadband (BI) are significant. Consistent with the hypothesis, the fourth attitudinal construct (service quality) significantly explained the behavioural intention to change service providers. The path from the overall attitudinal dimension to BI is also significant. As was hypothesised, paths from the primary influence and overall normative factors to BI are significant. The second normative construct (secondary influences) is significantly related to the behavioural intention to change service providers. Of the three control constructs two—self-efficacy and facilitating conditions resources—are significantly related to BI. However, the path from the third control construct (i.e., knowledge) to BI is not significant. The overall control factors also significantly explained BI. This means that all three dimensions of the determinants of BI (i.e., overall attitudinal, normative, and control construct) are significantly related to BI. Finally, both BI and the facilitating conditions resources are significant determinants of the actual behaviour of adopting broadband.

It is not possible to compare the predictability of broadband adoption model with Oh et al.'s (2003) study. This is because the aforesaid two studies have examined different independent and dependent constructs. For example, this study has employed BI and actual behaviour as dependent constructs, but in Oh et al. (2003) the ultimate dependent construct was attitude. However, the predictive power of the broadband adoption model can be compared to guiding models such as the TAM, TPB and DTPB. This is because constructs such as BI and behaviour and structure of the broadband adoption model are similar to the TAM, TPB, and DTPB.

Table 8.2 illustrates the comparison of previous studies for the adjusted R^2 obtained for both behavioural intention and actual behaviour. The comparison clearly demonstrates that the broadband adoption model performed as well as the previous studies. With regards to the behavioural intention value of the adjusted R^2 varied between 0.20 (Gefen & Straub, 2000) and 0.57 (Taylor & Todd, 1995) (Table 8.2), the adjusted R^2 for this study is found to be 0.43 (Table 6.17), which suggests the appropriate level of explained variance. This means that the independent variables considered in this study are important for understanding a consumer's behavioural intention to adopt broadband. In terms of behaviour, the adjusted R^2 reported in previous studies varied from 0.32 (Davis et al., 1989) to 0.51 (Davis, 1989) (Table 8.2). Since the adjusted R^2 value for this study revealed the variance in behaviour to be 0.40 (Tables 6.17 and 8.2), it falls within the acceptable range.

Table 8.2. Comparison of intention and behaviour in terms of adjusted R^2

Study	Theory	Adjusted R^2	
		Behavioural Intention	Behaviour
Davis et al., (1989)	TAM	---	0.45
Davis et al., (1989)	TRA	---	0.32
Davis (1989)	TAM	---	0.51
Taylor and Todd (1995)	DTP	0.57	0.34
Taylor and Todd (1995)	TPB	0.57	0.34
Taylor and Todd (1995)	TAM	0.52	0.34
Karahanna et al., (1999)	TRA + TAM	0.38	---
Agarwal & Karahanna (2000)	TAM & Cognitive Absorption	0.50	---
Gefen & Straub (2000)	TAM	0.20	---
Brown et al., (2002)	TAM	0.52	---
Koufaris (2002)	TAM + Flow Theory	0.54	---
Current Study	**TPB + DTPB + MATH**	**0.43**	**0.40**
Recommended level (Straub et al., 2004)	**---**	**0.40 or above**	**0.40 or above**

Figure 8.1. MBA illustrating overall impact of attitudinal, normative and control factors

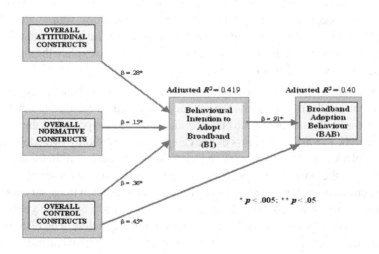

Therefore, similar to behavioural intention, behaviour also sufficiently explains the variance in broadband adoption by household consumers. The adjusted R^2 value for both the behavioural intention and behaviour also satisfy the criteria for predictive ability (Straub et al., 2004). Straub et al. (2004) suggested that the predictive ability of a model is satisfactory if the explained variance falls in the 0.40 range or above. Since both the values (BI = 0.43, B = 0.40) are within the range of 0.40, it suggests that the model possesses a satisfactory level of predictive ability.

Figure 8.1 depicts the overall paths from attitudinal, normative and control constructs towards the behavioural intention to adopt broadband. Consistent with the hypotheses, the overall attitudinal, normative and control constructs significantly explained the BI. This means that all three types of the determinants of BI (i.e., the overall attitudinal, normative and control constructs) have significant influence on BI. From the three types of determinants, the largest variance of BI was explained by the overall control factors, which was followed by the overall attitudinal factors and the overall normative constructs that explained the least variance of BI. Finally, both BI and facilitating conditions resources are significant determinants of the actual behaviour of adopting broadband.

Demographics and Adoption of Broadband

All of the five hypotheses, except one that was formulated to confirm the role of the socio-economic attributes, were supported by the collated data and chi-square test (Table 8.1). Of the five hypotheses, four were significant (at 0.05 level) and one was non-significant (Table 8.1). Therefore, the findings from the chi-square test are in line with the assumptions made whilst proposing the research hypotheses. A detailed discussion on the theoretical basis for each hypothesis and the expected outcome are provided in Chapter 2. Briefly, Figure 8.2 and Table 8.1 illustrate that of the five socio-economic characteristics selected for this study, four—age, education, income and occupation—significantly differentiated the broadband adopters from the non-adopters. However, the variable gender failed to explain the significant differences between the broadband adopters and non-adopters.

Early predictions of the impact of age on consumers and broadband adoption correspond to the results of this study. Earlier anecdotal evidence suggested that older people are less likely to subscribe to broadband, which was supported by the findings of this research. The above 65 years of age category achieved only two adopters but 12 non-adopters. A possible explanation is that the respondents within this category do not possess the basic skills to operate computers and the majority of them do not have computers at home because they do not consider them necessary for their needs. A majority of the adopters belonged to the age group of 25-54 years. This is because this age group is considered to be economically active. The respondents within this age group are mainly entrepreneurs, are in employment and are expected to have a high disposable income (Rice, 1997). Therefore, a new innovation is more likely to be adopted and diffused within this segment of the age groups (Rogers, 1995). A high number of non-adopters also belong to the age ranges of 25-54 years. A possible explanation of this is the lack of compelling content with a broadband connection. Although the members of this age group are economically active and possess high disposable incomes, they are reluctant to subscribe to broadband as they are of the opinion that broadband does not offer substantial advantages, or any added value in comparison to dial up.

In the case of gender, the proposed hypothesis was not supported by the data collected in this study. Although the adopters were more from the male category, whilst the non-adopters were more from female, the differences were not large enough to suggest the occurrence of significance. This may be because within the household, the choice of purchasing or subscribing to a service that requires financial commitment, that is, paying a monthly subscription fee, requires joint decisions from a majority of the households. Therefore, in household terms, gender may be less significant in explaining the differences between adopters and non-adopters of broadband. This theoretical claim was also supported by previous studies that reported decreases in the gender gap in terms of the computer and Internet access (Carveth & Kretchmer, 2002; Mason & Hacker, 1998). A study by Carveth and Kretchmer (2002) also provided similar indications of Internet users in the USA. This study suggested that in the USA, there are approximately equal numbers of men and women using the Internet (Carveth & Kretchmer, 2002).

It was expected that a large number of the educated respondents are more likely to adopt broadband. The findings of this are in accordance with the predictions for demographic variables, such as income and education. The findings suggest that the lowest proportion of the adopters possess only a GCSE qualification, which is the lowest listed level of education. In comparison, the majority of the adopters possessed an undergraduate and postgraduate level education. This is because broadband is a utility tool for accessing study material and an effective communication medium. Furthermore, highly educated people are most likely to hold higher occupation positions of employment; hence, they may need broadband to undertake office work at home.

The findings suggest that the majority of broadband adopters belong to either of the two higher occupational categories, 'A' or 'B.' It was not expected that the respondents from the lowest occupational category, that is, 'E,' would have a broadband connection. However, the findings imply that the largest number of broadband adopters belong to this latter category. This can be attributed to the following reasons: the respondents are in employment that maps to the occupational category 'E;' however other family members may hold higher levels of occupation. Second, as Rice (1997) argued, an anomaly in the occupational category may occur when a respondent who belongs to 'B' obtains redundancy in employment and then immediately drops to

Figure 8.2. Effects of demographic variables on broadband adoption

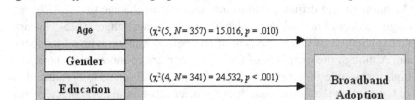

the grade 'E 'section. This is despite the fact that the disposable income could have increased due to redundancy payments (Rice, 1997).

Freeman (1995) also discusses the link between unemployment and the diffusion of ICTs. It was found in this research that the unemployed respondents in category 'E' are engaged in re-skilling in order to achieve white-collar jobs (Freeman, 1995). ICTs facilitate the process of re-skilling; therefore, it is more likely that such unemployed respondents become the adopters of broadband than non-adopters. The other reason behind this exception is that the occupational segment 'E' consists of a majority of respondents who are studying at various levels; thereby, they require Internet access at home. This is in accordance with Freeman's (1995) view that the most intensive use of technological resources such as computers comes after school hours (Freeman, 1995). Therefore, students are most likely to adopt new technologies including computers and broadband as a means of facilitating their studies and improving performance.

The findings also revealed that income levels are good predictors of broadband adopters and non-adopters, which is in line with the arguments offered in the theoretical section (Rogers, 1995; Venkatesh et al., 2000). Further, it can be learnt that the adopters are few from the lower income groups. However, the numbers of adopters increase as the income level rises. Therefore, the numbers of non-adopters are higher in the lower income group but decrease as the income level increases. The numbers of non-adopters is minimal in the highest income category.

Usage of Broadband

As described in Chapter 1, studies relating to the usage and impact of broadband have taken the form of user surveys, which have examined broadband users' behaviour in comparison to narrowband users. Results from initial studies suggested that Internet users behave differently when they have broadband access. Broadband users use the online facilities on a longer basis, utilise more services or applications and apply them more often (Anderson et al., 2002; Carriere et al., 2000; Dwivedi & Choudrie, 2003; Horrigan et al.,2001). In comparison to the dial up users, broadband users spend more total time on electronic media applications (Bouvard & Kurtzman, 2001).

Although these studies examined the usage of broadband, they lack theoretical underpinnings as they are data-led and exploratory in nature. In order to examine and confirm differences with regards to the usage of the Internet between the broadband and narrowband household consumers in the UK, this study employed two theoretical constructs; namely the rate of Internet use and variety of internet use (Shih & Venkatesh, 2004). Figure 8.3 illustrates that both the theoretical constructs successfully distinguished broadband consumers from the narrowband consumers. In terms of both duration of internet access (that is, how long consumers spend online) and frequency (that is how many times consumers access the internet on a daily basis), the broadband consumers significantly exceeded the narrowband consumers (Figure 8.3). This implies that due to the advantages that broadband offers, such as faster access, always-on access, faster download, un-metered access, consumers have been using the Internet in a more intense manner than the narrowband consumers.

Similar to the rate of use, the advantages of broadband also significantly influenced UK consumers in terms of the variety of Internet use (Figure 8.3). Variety in this context means the types of online services and/or accessed applications. In this study, a total of 41 online services that belonged to nine different categories were included to examine the variety of Internet use (Table 6.33). As described in Chapter 6, these

Figure 8.3. Usage of Internet by broadband adopters and non-adopters

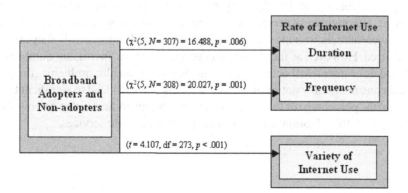

nine categories of online services and applications comprised of communications (five online services), information seeking (seven online services), information producing (four online services), downloading (six services), media streaming (five services), e-commerce (eight services), entertainment activities (four services), social and personal (two services), and e-government (Table 6.33).

This cross-sectional study confirmed that of the 41 activities that belonged to the nine different categories, broadband consumers on average used 22.41 activities, which significantly exceeded the narrowband users who utilised an average of 17 online activities from home. This is in line with the theoretical argument that the intense usage of technology leads to a higher variety of use (Shih & Venkatesh, 2004). Since broadband consumers used the Internet more intensely; they accessed a significantly larger number (i.e., more variety) of online services and applications than their narrowband counterparts.

The speed of the Internet was theoretically considered to be one of the barriers of growth and diffusion of electronic services, including electronic commerce; subsequently, they contributed to the doom of business-to-consumer electronic commerce (Rose et al., 1999). This is also supported by the findings of this research, as both the rate and variety of Internet use are less amongst the narrowband consumers. This means that the narrowband Internet connection hampers the growth and diffusion of emerging electronic services.

This is supported by the findings presented in Table 5.40, which illustrates that of the 41 activities, 20 online activities were accessed or utilised by significantly more broadband consumers than narrowband consumers.

The percentage of the consumers who accessed the remaining 21 activities were also more from the broadband categories than narrowband ones; however, the differences were not large enough to be significant. The 21 activities can be placed in the following two categories. First, online services such as e-mail, which does not require high speed access; hence, the services are used equally by narrowband and broadband consumers. Second, new online services, such as video conferencing and VoIP, which requires high speed Internet; hence, it is not convenient to access them utilising narrowband, and consequently, few respondents tried using such services. In contrast, although these services can be accessed utilising broadband, there are many new consumers who may not be aware of them and subsequently they are not being utilised. However, as these emerging services mature, increasing numbers of broadband consumers will access them, but due to bandwidth problems, the narrowband consumers may not be able to do so. This will then lead to significant differences in the percentage of the consumers which, again, indicates that narrowband will slow down the adoption and diffusion of new electronic services.

Impact of Broadband

According to the diffusion literature, new innovations are likely to change the associated behaviours of users, which are termed as perceived consequences or the impact of new innovations (Rogers, 1995; Shih & Venkatesh, 2004; Vitalari et al., 1985). Previous studies have demonstrated the impacts of various technologies (e.g. ,automobiles, telephones, computers and Internet) on a user's daily life (Anderson & Tracey, 2001; Vitalari et al., 1985). It was argued in Chapter 2 that since broadband offers an alternative way of work and entertainment and consumes time that traditionally has been spent on other activities, it is likely that broadband will alter the time allocation pattern of a user's daily activities. Therefore, it was also a part of the research aim of this study to investigate the impact of broadband upon changes in the time allocation patterns of consumers within UK households.

This discussion illustrates that the Internet usage behaviour of broadband consumers differs from the narrowband ones; hence, it is likely to have an impact on their daily life. The findings of this study indicate that the relationship between technology use and changes in time allocation patterns, which is depicted in Figure 8.4. The significant increase in total time spent online (Figure 8.3) and frequency of Internet access (Figure 8.3) may have triggered an imbalance in the equilibrium in the household systems that existed before the adoption and use of broadband. This imbalance is adjusted due to the changes in time allocation patterns on various daily life activities. The findings provide evidence that all 20 activities are affected by the change in time allocation patterns (Table 6.34). However, the magnitude of the effect was more on the activities closely related to broadband use. These include an increase in working at home and a decrease in working at an office, thereby decreasing the commuting in traffic. Similarly, watching television and reading were considered to be cognate activities and their usage patterns were closely similar to computer and Internet use (Vitalari et al., 1985). Therefore, a substantial number of respondents reported a reduction or increase in time for these activities (Table 6.34). It was also found that there is a significant difference between the narrowband and broadband consumers in terms of changes in time allocation patterns for five activities that comprised a decrease in shopping in store, increases in job related work at home, decreases in working in the office, decreases in commuting in traffic and a reduction in reading newspapers and books (Table 6.34).

Although differences occurred for the remaining 15 activities, they were not large enough to be statistically significant (Table 6.34). This was because the aforementioned five activities have a close relation to high-speed Internet connection use, whereas the other 15 do not. For example, performing job related tasks at home requires high-speed Internet service for efficiency and effectiveness reasons. Similarly, shopping online requires a high-speed Internet service for security and convenience reasons. Therefore, these activities were significantly affected by Internet use, as

Figure 8.4. Broadband impact in the household (An adaptation of homeostatic model of the effect of computer use on household time allocation patterns; Vitalari et al., 1985)

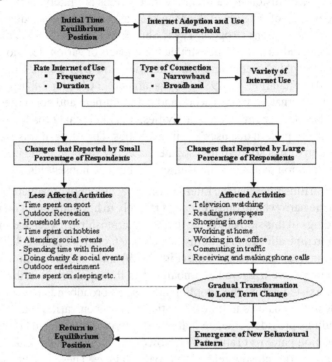

opposed to the impacts on other activities such as the time spent on sport, which has the least relation to Internet use.

This discussion supports the theoretical arguments presented in Chapter 2. The findings support the model of homeostasis that was utilised to conceptualise the impact of broadband on the time allocation patterns of households. This is similar to the previous study (Vitalari et al., 1985),which examined the impact of computer use in households. Subscribing to the Internet at home may affect a consumer in one of the following three possible ways.

First, a lack of change in the existing pattern of behaviour. As is illustrated in Table 6.34, respondents reported no change in the way they acted. This is because the characteristics of household users, such as age, are such that they elicit a low level of interest in the Internet (Vitalari et al., 1985).

Second, short term changes in the behaviour for those activities that are reported by a small number of respondents. Such activities are mainly external in nature and include sports, eating, and socialising with friends (Vitalari et al., 1985). The reason why a technology—in this case broadband—affects unrelated activities is that at the initial stage of adoption, users try to experiment and undertake new activities with technology and consequently have less time to spend on other activities. However, such changes slowly disappear as users get acquainted and adjusted to the use of new technology (Vitalari et al., 1985). It was also argued that the short-term changes might become long-term if the use of the Internet is continued over time (Vitalari et al., 1985). For example, online shopping may be an initial instance of a short-term change in consumer behaviour however, if there is continuous use of the Internet and a consumer develops trust and routine, then it may become a long-term change in household behaviour. However, it is not possible to demonstrate this within this study as the data is collected at one point of time. This requires a longitudinal approach to examine the impact of new technology, which cannot be achieved within this study due to the time constraints.

Third, long term changes in the household's behaviour for activities that are reported by a fairly large number of respondents. The affected activities that may bring long-term changes in behaviour are internal in nature and closely associated to the technology in question. Examples include watching television, reading, studying, increase in working at home, and decrease in working at an office (Vitalari et al., 1985). Activities such as working at home, working in the office, and commuting in traffic are interconnected and more likely to be affected by being connected using broadband, as it offers the opportunity to undertake office work at home.

This discussion indicates that household consumers will gradually reflect new behavioural patterns, where consumers will not spend their time visiting stores or shopping malls for the purchase of household commodities. Likewise, consumers may prefer utilising e-government services for obtaining pensions, benefits and paying taxes instead of visiting the council office or other public sector organisations. Gradually, an increasing number of people will begin working from home if their nature of work will allow it. This may benefit an organisation with cost savings and the workers with time savings, thereby providing better convenience and comfort. Several such issues may arise due to the use of broadband in households. The adoption and use of broadband may gradually affect the value and supply chains in several industries including media, as it is indicated from the findings that consumers prefer electronic media. Therefore, in the future, new electronic media may take the place of traditional media. For instance, online news may replace reading hard copy newspapers. As the Internet is gradually adopted and integrated in daily life and society, such issues will become visible; hence, future research may then be directed to investigate such issues individually and in an in-depth manner.

Summary

This chapter discussed and reflected upon the findings from the theoretical perspective. First, this chapter presented the refined and validated conceptual model of broadband adoption. The discussion led to the conclusion that all the constructs, except for knowledge, significantly explained the behavioural intention to adopt broadband; which in turn significantly explained the actual broadband adoption behaviour. A comparison of the adjusted R^2 obtained in this study with the previous studies suggests that the performance of the conceptual model that was used to understand the behavioural intention for the adoption and actual broadband adoption is as good as its guiding models.

Second, this chapter also discussed the usage and impact of broadband on household consumers. The discussion revealed that the broadband users differ from narrowband consumers with regards to the usage and impact of the Internet. From the discussion, it can be concluded that broadband consumers significantly differ from narrowband ones in terms of duration and the frequency of Internet access on a daily basis. It was also found that the use of broadband significantly affected the time allocation patterns on those activities whose execution is dependent upon the faster access of the Internet.

Chapter 11 will conclude the UK case study. However, before concluding the UK case study, Chapter 9 will examine the impact of broadband on the awareness and adoption of emerging e-government services. Following this, Chapter 10 will provide an empirical account of the service quality aspect of broadband subscription in the UK.

References

Agarwal, R., & Karahanna, E. (2000). Time flies when you're having fun: Cognitive absorption and beliefs about information technology usage. *MIS Quarterly, 24*(4), 665-694.

Ajzen, I. (1985). From intentions to actions: A theory of planned behaviour. In J. Kuhl & J. Beckmann (Eds.). *Action control: From cognition to behavior* (pp. 11-39). Heidelberg: Springer.

Ajzen, I. (1991). The theory of planned behaviour. *Organisational behaviour and human decision processes, 50*, 179-211.

Anderson, T. (2000). Regulation part 2: Digital transactions area a cause for concern. Retrieved September 15, 2004, from *Net imperative*. http://www.netimperative. com/2000/05/04/Regulation_Part_2

Anderson, B., Gale, C., Jones, M. L. R., & McWilliam, A. (2002). Domesticating broadband-what consumers really do with flat rate, always-on and fast Internet access? *BT Technology Journal, 20*(1), 103-114.

Anderson, B., & Tracey, K. (2001). Digital living: The impact (or otherwise) of the internet on everyday life. *American Behavioral Scientist, 45*, 456-475.

Bhattacharjee, S., Gopal, R.D., & Sanders, G.L. (2003). Digital music and online sharing: software piracy 2.0? *Communications of the ACM, 46*(7), 107-111.

Bouvard, P., & Kurtzman, W. (2001). *The broadband revolution: How super fast Internet access changes media habits in American households*. Retrieved October 25, 2002, from http://www.arbitron.com/downloads/broadband.pdf

Brown, S. A., Massey, A. P., Montoya-Weiss, M. M., & Burkman, J. R. (2002). Do I really have to? User acceptance of mandated technology. *European Journal of Information Systems, 11*(4), 267-282.

BSG Briefing Paper (2004). *The impact of broadband-enabled ICT, content, applications and services on the UK economy and society to 2010*. London. Retrieved November 15, 2004, from http://www.broadbanduk.org/news/news_pdfs/ Sept%202004/BSG_Phase_2_BB_Impact_BackgroundPaper_Sept04(1).pdf

Carriere, R., Rose, J., Sirois, L., Turcotte, N., & Christian, Z. (2000). Broadband changes everything. *McKinsey & Company*, Retrieved November 30, 2002, from http://www.mckinsey.de/_downloads/knowmatters/telecommunications/ broadband_changes.pdf

Carveth, R. and Kretchmer, S. B. (2002). The digital divide in Western Europe: Problems and prospects. *Informing Science, 5*(3), 239-249.

Choudrie, J., & Lee, H. (2004). Broadband development in South Korea: Institutional and cultural factor. *European Journal of Information Systems, 13*(2), 103-114.

Cowen, T. (2004). *Why the music industry is suing you, your neighbour, or your child*. The Social Affairs Unit Report, London. Retrieved April 15, 2005, from http://www.socialaffairsunit.org.uk/blog/archives/000183.php

Davis, F. D. (1989). Perceived usefulness, perceived ease of use, and user acceptance of information technology. *MIS Quarterly, 13*, 319-340.

Davis, F. D., Bagozzi, R. P., & Warshaw, P. R. (1989). User acceptance of computer technology: a comparison of two theoretical models. *Management Science, 35*(8), 982-1003.

DeLone, W. H., & McLean, E. R. (1992). Information systems success: The quest for the dependent variable. *Information Systems Research, 3*(1), 60-95.

DeLone, W. H., & McLean, E. R. (2003). The DeLone and McLean model of information systems success: A ten-year update. *Journal of Management Information Systems, 19*(4), 9-30.

Dwivedi, Y. K., & Choudrie, J. (2003). The impact of broadband on the consumer online habit and usage of Internet activities. In M. Levy et al. (Eds.), *Proceedings of the 8th UKAIS Annual Conference on Co-ordination and Co-opetition: the IS role*. Warwick, UK.

Dwivedi, Y. K., Choudrie, J., & Gopal, U. (2003). Broadband stakeholders analysis: ISPs perspective. In R. Cooper et al. (Eds.), *Proceedings of the ITS Asia- Australasian Regional Conference*, Perth, Australia.

Fishbein, M., & Ajzen, I. (1975). *Belief, attitude, intention, and behavior: An introduction to theory and research*. Reading: Addison-Wesley.

Freeman, C. (1995). *Unemployment and the diffusion of information technologies: the two-edged nature of technical change*. Policy Paper No. 32, ESRC.

Gefen, D., & Straub, D. W. (2000). The relative importance of perceived ease of use in IS adoption: A study of e-commerce adoption. *Journal of the Association for Information Systems, 1*, 1-28.

Heijden, H. (2004). User acceptance of hedonic information systems. *MIS Quarterly, 28*(4), 695-705.

Horrigan, J. B., & Rainie, L. (2002). *The broadband difference: How online Americans' behaviour changes with high-speed Internet connections at home*. Retrieved September 20, 2003, from http://www.pewinternet.org/pdfs/PIP_Broadband_Report.pdf

Karahanna, E., Straub, D. W., & Chervany, N. L. (1999). Information technology adoption across time: A cross-sectional comparison of pre-adoption and post-adoption beliefs. *MIS Quarterly, 23*(2), 183-213.

Koufaris, M. (2002). Applying the technology acceptance model and flow theory to online consumer behavior. *Information Systems Research, 13*(2), 205-223.

Lee, H., & Choudrie, J. (2002). *Investigating broadband technology deployment in South Korea*. Brunel- DTI International Technology Services Mission to South Korea. DISC, Brunel University, Uxbridge, UK.

Lee, H., O'Keefe, B., & Yun, K. (2003). The growth of broadband and electronic commerce in South Korea: contributing factors. *The Information Society, 19*, 81-93.

Mason, S. M., & Hacker, K. L. (2003). Applying communication theory to digital divide research. *IT & Society, 1*(5), 40-55.

Mathieson, K. (1991). Predicting user intentions: comparing the technology acceptance model with the theory of planned behaviour. *Information Systems Research, 2*(3), 173-191.

Mlcakova, A., & Whitley, E. A. (2004). Configuring peer-to-peer software: an empirical study of how users react to the regulatory features of software. *European Journal of Information Systems, 13*(1), 95-102.

Moore, G. C., & Benbasat, I. (1991). Development of an instrument to measure the perceptions of adopting an information technology innovation. *Information Systems Research, 2*(3), 192-222.

Oh, S., Ahn, J., & Kim, B. (2003). Adoption of broadband Internet in Korea: the role of experience in building attitude. *Journal of Information Technology, 18*(4), 267-280.

Premkumar, G. P. (2003). Alternative distribution strategies for digital music. *Communications of the ACM, 46*(9), 89-95.

Rose, G., Khoo, H., & Straub, D. W. (1999). Current technological impediments to B-2-C Electronic Commerce. *Communications of the Association for Information Systems, 1*(16). Reprint accessed from http://cais.isworld.org/articles/1-16/article.htm

Rice, C. (1997). *Understanding customers.* Oxford: Butter worth-Heinemann.

Rice, R. E., Grant, A. E., Schmitz, J., & Torobin, J. (1990). Individual and network influences on the adoption and perceived outcomes of electronic messaging. *Social Networks, 12*(1), 27-55.

Rogers, E. M. (1995). *Diffusion of innovations.* New York: Free Press.

Rosemann, M., & Vessey, I. (2005, June 26-28). Linking theory and practice: Performing a reality check on a model of IS success. In *Proceedings of the 13th European Conference on Information Systems*, Regensberg, Germany.

Shih, C. F., & Venkatesh, A. (2004). Beyond adoption: Development and application of a use-diffusion model. *Journal of Marketing, 68*, 59-72.

Straub, D. W., Boudreau, M-C, & Gefen, D. (2004). Validation guidelines for IS positivist research. *Communications of the Association for Information Systems, 13*, 380-427.

Tan, M., & Teo, T. S. H. (2000). Factors influencing the adoption of Internet banking. *Journal of the Association for the Information Systems, 1*.

Taylor, S., & Todd, P. A. (1995). Understanding information technology usage: A test of competing models. *Information Systems Research, 6*(1), 44-176.

Tornatzky, L. G., & Klein, K. J. (1982). Innovation characteristics and innovation adoption-implementation: A meta-analysis of findings. *IEEE Transactions on Engineering Management, 29*, 28-45.

Venkatesh, V., & Brown, S. (2001). A longitudinal investigation of personal computers in homes: Adoption determinants and emerging challenges. *MIS Quarterly, 25*(1), 71-102.

Venkatesh, A., Shih, C. F. E., & Stolzoff, N. C. (2000). A longitudinal analysis of computing in the home census data 1984-1997. In A. Sloane & F. van Rijn (Ed.), *Home informatics and telematics: Information, technology and society* (pp. 205-215). Norwell, MA: Kluwer Academic Publisher.

Vitalari, N. P., Venkatesh, A., & Gronhaug, K. (1985). Computing in the home: shifts in the time allocation patterns of households. *Communications of the ACM, 28*(5), 512-522.

Chapter IX

Exploring the Role of Broadband Adoption and Socio-Economic Characteristics in the Diffusion of Emerging E-Government Services

Abstract

The penetration of the Internet and opportunities involving information and communication technologies (ICTs) have occurred at an escalating rate within the private sector. This has caused governments and public sector organisations around the globe to become aware of their potential and consequently utilise them; thereby triggering investments in e-services (Choudrie et al., 2004a). However, the e-services offered by governments are much more than simple automation. E-services are meant to dramatically improve all areas of government activities: from democratic participation using online voting to improving the efficiency of citizen interactions with the government by providing online government services (Barc & Cordella 2004). Other countries around the globe are undertaking various e-government measures. Similar to them, the United Kingdom (UK) government is also undertaking steps that ensure it is progressing in accordance with these countries. For this purpose, the introduction and experiences of utilising a number of government services that are offered electronically via the 'Government Gateway' in the UK is pertinent. These e-government services include council tax bills and accounts, housing benefits, child benefit claims, carer's allowance, jobs online, state pension forecasts, self assessment of tax and tax credits, landweb direct, and the national blood services (Government Gateway, 2004). This is in contrast to the traditional method, whereby access to government services was undertaken by the citizens visiting 'physical, brick foundation' locations. Therefore this novel phenomenon offers a great amount of convenience and ease for citizens' use (Government Gateway, 2004). However, it is not known whether the citizens of the UK are aware of such services. Further, questions are still emerging concerning whether the citizens are actually adopting the newly offered services. Therefore the initial aim of this chapter is to examine citizens' awareness and adoption of e-government initiatives, specifically the Government Gateway in the United Kingdom. Since these services have been recently introduced, an investigation is needed to study if the demographic characteristics and home Internet access are affecting the awareness and adoption of these services. Therefore the second aim of this chapter is to examine the affect of the citizens' demographic characteristics and home Internet access on the awareness and adoption of e-government services. To fulfil these aims, this study undertook an empirical examination of the awareness and adoption of the Government Gateway amongst UK citizens. This research offers a contribution to various stakeholders including the government agencies who could require a distinction to be drawn between the adopters and non-adopters of e-government services. That is, from the results of this research the government agencies could better understand in a simpler and detailed manner, the problem of low adoption. This could allow the formulation of a strategy that promotes awareness and diffusion. The chapter begins with a brief discussion of research undertaken on the citizens' adoption of e-government services and a brief overview of the Government Gateway and its purpose. The findings are then presented and discussed. Finally, a conclusion to the research is provided. It is important to mention that the research methodology for this chapter is already presented and discussed in previous chapters. Therefore, this chapter does not include research methodology and only presents empirical data in the form of tables.

Background

The Definition and Benefits of E-Government Services

E-government is being defined in various ways, each differing according to the afforded purpose. In this research, the definition being applied is the one offered by the European Information Society:

E-government is the use of information and communication technology in public administrations combined with organizational change and new skills in order to improve public services, democratic processes and strengthen support to the public policies. (European Information Society, 2004)

According to the European Information Society (EIS) (2004), if appropriate e-government services are developed and delivered, then enormous benefits will be offered to all the stakeholders including the citizens, government, and businesses. For example, a recent survey of the European Union countries was conducted and it was found that the most frequently cited benefit of obtaining e-government services for the citizens, or in this case the users, are time savings and gaining flexibility. The service that provided citizens with greatest satisfaction was the ease of use offered by the centralised portals (EC Press Release, 2003). The survey study concluded that the crucial elements for the success of online government services are optimising workflows, simplifying processes, and improving the way that information is re-used and shared amongst the public authorities (European Information Society, 2004).

The EIS emphasised that the correct implementation of e-government will result in shorter queues for citizens in government offices. Added benefits of the improved service include: (1) reductions in the costs for both businesses and governments by cutting the tax burden and boosting competitiveness; (2) with the help of e-government the public sector can be made more open and transparent, delivering governments that are more comprehensible and accountable to citizens, improving civic involvement in policy making and reinforcing democracy at every level across Europe; and (3) administrations can be made more citizen-centred and inclusive, thereby providing 24/7 personalised services to everyone no matter what their circumstances or their special needs are (European Information Society, 2004).

To realise the overall benefits in Europe, the European Commission established the e-Europe 2005 target. *e*-Europe 2005 applied a number of measures to address simultaneously both the demand and supply sides of the equation. On the demand side, actions on e-government, e-health, e-learning, and e-business are designed to foster the development of new services. Additionally, in order to provide both better and cheaper services to citizens, public authorities can use their purchasing power to aggregate demand and provide a crucial pull for new networks. On the supply side, actions on broadband and security should advance the rollout of the infrastructure (e-Europe Action Plan Report, 2002). The next section provides a brief review of the work focused upon citizens' adoption of e-government services in the UK.

Citizen Adoption of E-Government Services

In the adoption pattern area, the topic of e-government is a new and emerging one. Due to this, research in this area has focused upon the supply side or government related issues such as strategies and policy (Beynon-Davies 2004; Chadwick & May, 2003; Choudrie et al., 2004b; Williams & Beynon-Davies, 2004), challenges (Barc & Cordella, 2004; Weerakkody et al., 2004); technical issues (Cottam et al., 2004; George, 2004), and the evaluation of the usability of e-government websites (Choudrie et al., 2004; Mosse & Whitley, 2004). However, little attention has been given to the demand or citizen perspective. Recent studies that addressed the citizen adoption of e-government services suggest that trust and security (Otto, 2003) and transparency (Marche & McNiven, 2003) are the major issues for e-government adoption.

Gefen et al. (2002) studied the adoption of online government services employing data collected from the undergraduate students of three Universities. Findings of this study suggest that trust, social influence, and website ease-of-use impact perceived usefulness of the interface which, combined with social influence, predict the intended use of e-government. Carter and Belanger (2004a) integrated constructs from the Davis (1989) technology acceptance model (TAM), the Rogers (1995) diffusion of innovation theory (DOI), and Web trust (McKnight et al., 2002) research. Findings of this study suggest that the compatibility of relative advantage and perceived usefulness are constructs that are important in explaining citizens' perception of e-government adoption (Carter & Belanger, 2004a). Carter and Belanger's (2004b) study used Moore and Benbasat's (1991) perceived characteristics of the innovations' constructs to identify factors that influence the citizen adoption of e-government initiatives. Findings from this study reveal that relative advantage, and image and compatibility are the important constructs when predicting intentions to use the state e-government services (Carter & Belanger 2004a).

These studies were initial attempts to examine factors such as trust, transparency, compatibility, and relative advantage, which affect the adoption of e-government services by the citizens of the United States of America (USA). Literature analysis suggests that such studies have not yet been undertaken to investigate the adoption behaviour of UK citizens. This is because e-government-related development is new and recent to the UK. A report on benchmarking e-government in the Europe and the USA found that 'existing studies of e-government concentrate on the supply-side by focusing on the availability and level of sophistication of online services and usage' (Rand Europe, 2003). The findings from this study reveal that the citizens are reluctant to utilise e-government services that require personal information. These findings also suggest that 'among respondents who indicated a preference for online government services, citizens were not always aware of which government services were available online' (Rand Europe, 2003).

This report covered only a few aspects of the demand-side of e-government and recommended that 'future research is essential to get a more complete picture of the perceptions and attitudes of the users of e-government services' (Rand Europe, 2003). Therefore, bearing these reasons in mind, the aim of this research became to survey the state of awareness and adoption of the recently introduced e-government initiative 'Government Gateway' that offers a centralised registration for all e-government services available in the UK. The following subsection provides a brief discussion about the Government Gateway and the process that can be utilised to register on it.

The Development of E-Government Services in the UK: Government Gateway

Governments around the globe are developing and implementing e-government initiatives, with the UK government being no exception. A centralised e-service initiative that is of current interest and has been developed by the UK government is the 'Government Gateway.' This centralised portal enables citizens of the UK to access any of the online e-government services (Government Gateway, 2004). The UK government has made available a number of services that can be accessed by the citizens electronically via the common 'Government Gateway.' These e-government services range from paying the council tax to agricultural and rural development issues (Government Gateway, 2004). Some of the important e-government services that are currently available include: (1) the citizens of the Borough Council of King's Lynn and West Norfolk can view their council tax bills and accounts, and submit direct debit applications in order to pay a council tax.

Additionally, individuals can query the amount of entitled housing benefit and /or council tax benefits and submit various changes of circumstances; (2) farmers from Northern Ireland can submit online the sheep annual premium scheme application (SAPS); (3) services related to the Department for Environment, Food, and Rural Affairs (DEFRA) can be conducted online; (4) services from the Department for Work and Pensions (DWP) such as a child benefit claims, carer's allowance, online jobs, and state pension forecasts are available online; (5) Inland revenue services such as self assessment of tax and tax credits; (6) Landweb direct service for Northern Ireland's professionals dealing in the land and property business; and (7) the national blood services are currently accessible online in the UK. Contrasting this to the traditional method whereby access to government services required citizens visit different web addresses, the common Government Gateway offers a means to the various e-government services using a centralised path. Therefore this Gateway offers a great amount of convenience and ease for the e-citizens use (Government Gateway).

The Government Gateway registration and enrolment process involves undertaking the following steps: (1) select to register as an individual, organisation, or agent; (2) determine whether to receive a user ID or use a digital certificate; (3) enroll for government service(s); (4) receiving of activation PIN for each enrolled service through the post; and (5) activation of the service. After completing the registration process, a single user ID or digital certificate is allocated to the applicant and it can be used for all the Government Gateway services. Employing the allocated ID and PIN, transactions can be made using the appropriate government Websites, portals, or third party software packages. As explained previously, the process of registration and enrolment is simple and straightforward for users. However, as yet, the awareness of and registration to the centralised service amongst the citizens is not yet known. This chapter is an initial attempt to provide empirical evidence on the awareness and adoption of the Government Gateway by the citizens of the UK.

Findings

Citizens Awareness and Adoption of Government Gateway

Since the focus of this research is upon the awareness and adoption of e-government services, the next step to this research involved categorizing the percentage of the citizens into: (1) those who registered to the Government Gateway; (2) those aware of the Government Gatweay, but not registered; and (3) those who are not even aware of the Government Gateway. Table 9.3 illustrates these findings. The results indicate that only 6 percent of the respondents of this sample had registered. Of the remaining 94 percent respondents, 18 percent stated that they had not yet registered and the remaining 78 percent were not even aware of the Government Gateway (Table 9.3).

Table 9.3. Awareness and adoption of Government Gateway (N = 358)

	Yes	No
Government Gateway Awareness	24%	76%
Government Gateway Adoption	6%	94%

Respondents Age and Awareness and Adoption of Government Gateway

It was found that as increases in the citizens' age ranges occurred, there was more awareness of and adoption of the Government Gateway (Table 9.4). An interesting result was that the adoption rate declines considerably after 54 years and there were no respondents reported at the 64 years and above category. The majority of the adopters were between the age ranges of 25 and 54 years. The findings illustrated in Table 9.4 also suggest that although the younger citizens are aware of the Government Gateway, they are reluctant to register. Contrastingly, the older age group consisted of more respondents who were not aware of the Government Gateway. This may be because this age group consists of mostly non-computer users who do not possess the skills and knowledge necessary in order to use the computer and Internet.

Respondents Gender and Awareness and Adoption of Government Gateway

In terms of gender differences, Table 9.5 illustrates that in terms of both awareness and adoption, there are more males in comparison to females. This suggests that at

Table 9.4. Age and awareness and adoption of Government Gateway (N = 357)

	Gov. Gateway Awareness	Gov. Gateway Adoption
<=24	19%	10%
25-34	25%	5%
35-44	28%	33%
45-54	25%	43%
55-64	24%	10%
=>65	12%	0%

Table 9.5. Gender differences on awareness and adoption of Government Gateway (N = 355)

	Male	Female
Gov. Gateway Awareness	59%	41%
Gov. Gateway Adoption	76%	24%

initial stage of implementation of the e-government services males are more likely to drive adoption.

Respondents Education and Awareness and Adoption of Government Gateway

Table 9.6 offers an explanation of the educational attainment of the survey respondents. The findings suggest that the majority of adopters are educated to the postgraduate and degree levels. Although a small number of the adopters possess an education qualification below degree level, awareness does exist across all the segments of educational attainment. Hence, it is clearly apparent from the findings that education is an important demographic factor that will affect the growth and diffusion of future e-government services.

Table 9.6. Educational differences on awareness and adoption of Government Gateway (N = 341)

	Gov. Gateway Awareness	Gov. Gateway Adoption
GCSC	5%	0%
GNVQ/Diploma	6%	5%
Advance Level	11%	5%
Undergraduate	48%	48%
Postgraduate	43%	43%

Table 9.7. Occupational differences on awareness and adoption of Government Gateway (N = 345)

	Gov. Gateway Awareness	Gov. Gateway Adoption
A	16%	24%
B	36%	57%
C1	12%	5%
C2	4%	5%
D	1%	0%
E	31%	10%

Respondent Social Class and Awareness and Adoption of Government Gateway

The social classes were derived using the occupation listed responses. Mainstream professionals such as, doctors, lawyers, and judges with the responsibility of more then 25 staff belonged to social class 'A.' The occupations that had a responsibility of less then 25 staff were classified as category social class 'B' and included academics. Skilled-non-manual workers fell within social class 'C1' and 'C2.' The unskilled manual workers belonged to social class 'D.' Finally, social class 'E' consisted of the pensioners, casual workers, un-employed individuals, and students (Rice, 1997, p. 241).

Table 9.7 illustrates the occupational category or social class of respondents who are aware of and registered for the Government Gateway. This suggests that no respondents from social class 'C2' and 'D' provided responses. The findings suggest that the prevalent number of respondents who are aware of the Government Gateway belonged to social class 'E' which was then followed by class 'B.' Contrastingly, the prime number of adopters who registered at the common Gateway belonged to social class B, followed by social class 'A'. In general terms, it is suggested that the awareness and adoption of e-government services decreases as the social class of respondents reduces. This is in line with findings presented in previous chapters suggesting that the broadband adopters emanate from a higher rather than a lower social class.

Table 9.8. Income differences on awareness and adoption of Government Gateway

	Gov. Gateway Awareness	Gov. Gateway Adoption
<10K	7%	0%
10-19K	15%	5%
20-29K	19%	24%
30-39K	13%	14%
40-49K	10%	19%
50-59K	7%	5%
60-69K	15%	19%
=>70K	13%	14%

Respondents Income and Awareness and Adoption of Government Gateway

Table 9.8 illustrates that although almost all the income segments are aware of the Government Gateway, the adopters belonged mainly to the higher income group. For instance, 7 percent of the respondents from the income group £10-19 K were aware of the service. However, none of the respondents had registered to the Government Gateway. Generally, the higher the income of respondents the more probable it was that the e-government services would be adopted.

Broadband Access and Awareness and Adoption of Government Gateway

Table 9.9 illustrates the clear differences amongst those who do or do not have access to broadband at home. Respondents with broadband access at home had both higher awareness and adoption rates than those who did not. This suggests that the diffusion and adoption of broadband at home will directly affect the adoption and diffusion of new electronic services such as e-government services.

Conclusion

The aim of this research was to examine citizens' awareness and adoption of e-government initiatives, specifically the 'Government Gateway' in the UK. To fulfil this aim, an initial and exploratory study was undertaken utilising the Government Gateway,' a centralised registration facility of all available e-government services. Also examined within this chapter was the role of demographic variables and home Internet access in the awareness and adoption of e-government services. The study found that as with other similar ICTs such as computers, Internet and broadband, the demographic characteristics of citizens such as age, gender, education, and social

Table 9.9. Effect of broadband adoption on Government Gateway adoption (N = 308)

	Gov. Gateway Awareness	Gov. Gateway Adoption
Narrowband	21%	14%
Broadband	79%	86%

class play an imperative role in explaining the citizen's awareness and adoption of e-government services in the household. This study also concluded that citizens with home Internet access are more likely to be aware of and adopt e-government services, in this case, the 'Government Gateway.'

References

Barc, C., & Cordella, A. (2004). Seconds out, round two: Conceptualizing e-government projects within their institutional milieu-a London local authority case study. In *Proceedings of the 12ᵗʰ European Conference on Information Systems*, Turku, Finland.

Beynon-Davies, P. (2004). Constructing electronic government: The case of the UK Inland Revenue. In *Proceedings of the 12ᵗʰ European Conference on Information Systems*, Turku, Finland.

Carter, L., & Belanger, F. (2004a). Citizen adoption of electronic government initiatives. In *Proceeding of the 37ᵗʰ Hawaii International Conference on System Sciences*.

Carter, L., & Belanger, F. (2004b). The influence of perceived characteristics of innovating on e-government adoption. *Electronic Journal of E-government*, *2*(1), 11-20.

Chadwick, A., & May, C. (2003). Interaction between states and citizens in the age of the Internet: "E-government" in the United States, Britain, and the European Union. *Governance, 16*(2), 271-298.

Choudrie, J., Ghinea, G., & Weerakkody, V. (2004a). Evaluating global e-government sites: A view using Web diagnostic tools. *Electronic Journal of E-Government, 2*(2), 105-114.

Choudrie, J., Papazafeiropoulou, A., & Light, B. (2004b). E-government policies for broadband adoption: the case of the UK Government. In *Proceedings of the Americas Conference on Information Systems*, New York.

Cottam, I., Kawalek, P., & Shaw, D. (2004). A local government CRM maturity model: a component in the transformational change of the UK councils. In *Proceedings of the Americas Conference on Information Systems*, New York.

Davis, F. D. (1989). Perceived usefulness, perceived ease of use, and user acceptance of information technology. *MIS Quarterly, 13,* 319-340.

European Commission Press Release (2003). *Citizens and business welcome e-government services.* Retrieved July 14, 2005, from http://europa.eu.int/rapid/pressReleasesAction.do?reference=IP/03/1630&format=HTML&aged=0&language=EN&guiLanguage=en

European Information Society (2004). Retrieved July 14, 2005, from http://europa. eu.int/information_society/progrmmes/egov_rd/about_us/text_en.htm

e-Europe Action Plan Report (2002). *eEurope 2005: An information society for all.* Retrieved July 14, 2005, from http://europa.eu.int/information_society/eeurope/2002/news_library/documents/eeurope2005/eeurope2005_en.pdf

Gefen, D., Warkentin, M., Pavlou, P. A., & Rose, G. M. (2002). E-government adoption. In *Proceedings of the Americas Conference on Information Systems,* Texas.

George, K. (2004). Network analysis of disconnect in the hollow state: The case of e-government services portals. In *Proceedings of the 12th European Conference on Information Systems,* Turku, Finland.

Government Gateway Portal, Retrieved November 10, 2004, from http://www. gateway.gov.uk

Marche, S., & McNiven, J. D. (2003). E-government and e-governance: The future isn't what it used to be. *Canadian Journal of Administrative Sciences, 20*(1), 74-86.

McKnight, H., Choudrie, V., & Kacmar, C. (2002). Developing and validating trust measures for e-commerce: An integrative typology. *Information Systems Research, 13*(3), 334-359.

Mosse, B., & Whitley, E. A. (2004). Assessing UK e-government websites: classification and benchmarking. In *Proceedings of the 12th European Conference on Information Systems,* Turku, Finland.

Moore, G. C., & Benbasat, I. (1991). Development of an instrument to measure the perceptions of adopting an information technology innovation. *Information Systems Research, 2,* 192-222.

Otto, P. (2003). New focus in e-government: from security to trust. In *Proceedings of International Conference on E-Governance,* Tata McGraw Hill Publishing Company Limited, New Delhi, India

Rand Europe (2003). *Benchmarking e-government in Europe and the US.* Available at http://www.rand.org/publications/MR/MR1733/MR1733.pdf

Rice, C. (1997). *Understanding customers.* Oxford: Butterworth-Heinemann.

Rogers, E. M. (1995). *Diffusion of Innovations.* New York: Free Press.

Weerakkody, V., Choudrie, J., & Currie, W. (2004). Realizing e-government in the UK: Local and national challenges. In *Proceedings of the Americas Conference on Information Systems,* New York.

Williams, M. D., & Beynon-Davies, P. (2004). Implementing e-government in the UK: An analysis of local-level strategies. In *Proceedings of the Americas Conference on Information Systems,* New York.

Chapter 10

Broadband Quality Regulation:
Perspectives from UK Users

Elizabeth A. Enabulele, Brunel University, UK

Gheorghita Ghinea, Brunel University, UK

Abstract

Broadband is a technology still penetrating the telecommunication market and is seen as the most significant evolutionary step since the emergence of the Internet. However we argue that in the rush to achieve market share, insufficient attention has been paid to quality issues, the central theme of this chapter. Indeed, the concept of quality is a multi-faceted one, for which various perspectives can be distinguished. In this chapter, we take a look at broadband quality as perceived by users in the UK and report the result of a survey, which determined the users' perception on broadband quality. The results of the survey show, that quality, though desired by many, has been short-changed by the desire to have access to the Internet via broadband at the lowest cost possible. However, this has not encouraged some consumers to take on broadband access despite some low prices offered by service providers as these low prices for broadband access is not commensurate to their needs.

Introduction

Although the Internet boom of the late 1990s fuelled a proliferation of projects for Internet access, the bursting of the dotcom bubble in 2001 had a major impact on all these projects. This major blow was in part a result of the lack of regulation, quality control, and guidelines with policies in which the measurement of the quality of the delivery of service is compared and adhered. A frequent occurrence was that different dotcom services were operating based on their own technical know-how, but without proper quality guidelines to safeguard against failure. Broadband technology will face a similar fate to the dotcoms if issues, such as the quality offered by service providers to their customers, are not properly addressed and the right regulation not put in place. For example, in December, 2001, Excite@Home, then the USA's largest broadband service provider, switched off its high-speed network affecting 45 percent of USA households with broadband access and losing as much as $6 million a week because of onerous contracts with its cable partners (Howard, 2002). Other broadband service providers, such as Northpoint and Rhythms Netconnections, also failed and ceased to exist despite a healthy demand for broadband connection growing appeal. This failure was a result of companies wanting to do too much too soon, financial commitment (Howard, 2002) and lack of regulatory policies.

While the many variables necessary to create commercially viable broadband media services are in place, these including several technological, service level agreement (SLA) and legal factors, broadband is also affected by the same issues as other segments of the computer industry, for example privacy, security and encryption, and quality. However, the role of the operating system, bandwidth requirements, network optimisation, policies, quality, security, and privacy have become part of the huge concern in the United Kingdom and indeed throughout the European Union for the provision of broadband services. These issues need to be addressed and overcome by the government, telecommunications companies, and service providers in order to provide broadband quality to its consumers.

The focus of earlier broadband research has been on the distribution and adoption of broadband Internet access which depends primarily on the construction of a national-level infrastructure or on macro-level factors such as government policies, market competition and the density of population (Sangjo et al., 2003). However, this earlier research on broadband has all approached the issue of quality in a fragmented manner.

Accordingly, earlier work on broadband quality has been systems oriented (Marchetti, 2004), focusing on issues such as traffic analysis (Leung, 2002), scheduling (Chen & Lee, 2000), cost (Lorenz & Orda, 2002), routing (Orda, 1999), and recently user-level issues (Choudrie & Dwivedi, 2004). Indeed, most of the research, as the enumeration exemplified, has focused on quality in a fragmented manner and often been misinterpreted. For example, quality of service (QoS) in broadband has until now been assumed by many to be the same as quality in broadband. However, quality is made up of many facets, of which QoS is but one of them. In this chapter, a unified multidimensional broadband quality framework will be proposed, as to the best of our knowledge none yet exists, indubitably, relatively little work has been done looking at broadband quality issues from a unified perspective.

The current research differs from previous work and offers an advanced and forward-looking perspective to previous work by providing a holistic approach to broadband quality. The results of our research can help service providers and indeed the government to understand how to focus on what consumers' perspective on broadband quality in a unified form is rather than imposing high specification technology that is not necessarily desired by users. Accordingly, this chapter is structured as follows: section 2 presents our proposed integrated broadband quality framework, with the research methodology of our study being presented in section 3. Section 4 describes the research findings; lastly, conclusions are drawn and possibilities for future work identified in section 5.

An Integrated Broadband Quality Framework

Despite the widespread broadband deployment, the proliferation of broadband technology still faces significant hurdles, one of which is quality. However the concept of quality is a multi-faceted one, as there are as many quality perspectives as there are stakeholder groups (Choudrie et al., 2005). Moreover, Pirsig (1994) wrote that once there is an awareness of quality it becomes difficult to let go of it, which might be one of the reasons broadband quality has thus far been treated in a fragmented manner. A central issue in this chapter will be the one of quality and is one of the main concerns experienced in today's information technology (IT) systems. Emphasis on quality has emerged in a variety of organisation and in several fields, which includes the IT industry and has become a significant concept in the marketplace. Quality in the IT field is usually measured by product failure (Krishnan, 1993), however it also relates to the characteristics by which customers or stakeholders judge an organisation, product or service.

Whilst quality is a subjective concept for which each person has his/her own definition, in the broadband realm, this is manifested through the different perspectives that the stakeholders accept as true. In the past, freedom from problems with services was considered to be a positive concept because customers expected significant problems; presently quality focus is generally on process quality, the avoidance and elimination of flaws. It is certain that failure to conform to a standard will produce dissatisfaction on the part of customers. Therefore service providers and telecommunication companies now seek to produce services which will be chosen by the customer based on needs and service level agreement (SLA). Although customers expect the basic requirements of their broadband to be met, they do not gain a feeling of excitement when only these are achieved.

Proposed Integrated Quality Framework

Inevitably, our research suggested that the quality of broadband service offered by service providers does not only depend on QoS issues, but other aspects of quality, such as price, perception, customer service, cultural quality, reliability, policy, security, and functional quality (Figure 10.1). All these quality aspects can be fused together to provide an integrated quality framework for the stakeholders, which include users, service providers and government.

*   **Price:** Pricing is an important aspect of quality that needs to be looked at in broadband and can be a crucial limit on the demand for broadband (Romero, 2002). Pricing in broadband can be used as an effective means to increase competition among different service providers (Choudrie & Lee, 2004). From the customers and service providers' perspective, it is assumed that customers and potential customers of broadband services are most concerned with clear information about the service offering, particularly concerning the speed of download and upload that can be expected in practice, clear pricing information and activation process that is predictable.

Figure 10.1. Integrated quality framework

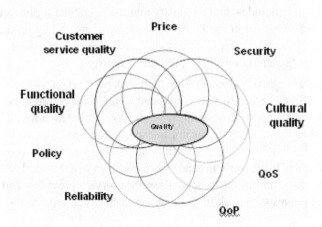

- **Security:** This is a potential issue to broadband adoptions particularly the increased exposure to security incidents and vulnerabilities. Examples of these are viruses, worms, and hackers; thus precautions are necessary. The highest priorities for broadband users to consider for their connection are firewall, anti-virus, and privacy control/content filtering capabilities. Firewalls screen data passing through the device, filtering out that which may be deemed harmful.

- **Cultural quality:** Cultural quality is an effort to improve the level of performance across an entire organisation to achieve higher levels of customer satisfactions. This is also referred to as continuous quality improvement (CQI). CQI is the use of incremental and breakthrough quality management techniques to constantly improve processes, products, or services provided to internal and external customers and thus achieve higher level of customer satisfaction.

- **Quality of service:** QoS is a key attribute of a broadband communication system (Koraski & Tassiulad, 2004). It is a set of methods and processes that a service provider implements to maintain a specific level of quality. It has been proposed to capture the qualitatively or quantitatively defined performance needed of an application. QoS broadband networks provide customers with the capability to select the kind of service they require and, therefore, obtain more efficient allocation of network resources.

- **Quality of perception (QoP):** To understand and predict the performance users will experience when they use the broadband, one needs a thorough understanding of the processes of human perception. In this respect, QoP is a metric that encompasses not only users' satisfaction with the quality of multimedia presentations but the users' ability to analyse, synthesise and assimilate the informational content of a multimedia display (Ghinea et al., 1999).

- **Reliability:** Service providers are expected to take appropriate steps for making required bandwidth available in a time bound manner within their licence framework. The cost of bandwidth constitutes a major cost component for broadband service (Kumar, 2004). Having a reliable connection provides users with a sense of quality that their broadband connection is stable.

- **Policy:** The long-term protection for provider of content lies in robust competition among providers of broadband connectivity (Huber, 2002). The Telecommunication Act of 1996 established a regulatory policy that promoted competition, innovation, and investment in broadband services and facilities. This helps in providing a good foundation for the provision of broadband service.

- **Functional quality:** This is the testing of the software that runs the broadband connection from the user to the service provider. In order to achieve this software quality it is expected that the necessary testing procedures and techniques has been utilised before deployment of such software to the users. This testing involves using black box testing, which gives a functional testing of the software. This procedure involves making sure that the broadband equipment works flawlessly with it and testing compatibility with various operating systems.

- **Customer service quality:** The emphasis on service quality has been most apparent since the growth of managed customer service. It forms the basis for clear communication with the customer with the objective of ensuring that customers are aware of what they are purchasing and the degree to which the available products will meet their needs and expectations. By clearly defining the terms that are used, customers can have greater confidence that their expectations are realistic.

A result of the fragmentation of broadband quality issues is also the fact that users are usually asked about their opinions with respect to broadband quality only through the prism of highly topical surveys, and never from a holistic perspective. In this chapter, we address this shortcoming and present the results of a pilot study which sought user opinions on the various aspects of broadband quality that our proposed integrated framework contains.

Research Methodology

As this is an initial exploratory study on broadband quality, the survey method was considered to be the most suitable research method for this investigation. Previous research has shown that the survey method is most appropriate when investigating technology adoption (Venkatesh & Brown, 2001).

Determining the Sample

The data for this research was obtained from a nationwide pilot survey of UK consumers, which included academicians, students, home users, and professionals in their respective jobs within a short period of time (September 2005-December 2005). This survey questionnaire was used to investigate users' perception on broadband quality as provided by their service providers. A deliberate effort was made to ensure that a wide variety of stakeholders were included in order to have a fair assessment on the pilot survey. The stakeholders selected were those individuals who use broadband in their office, homes or elsewhere and those that do not have access to broadband.

Questionnaire Development, Pilot Study and Data Collection

The survey questions addressed general and background information regarding each respondent however this excluded their names for security reasons. Furthermore, the survey questions focused on users' perception of 10 issues dealing with broadband quality based on ratings given using a 7-point scale ranging from *strongly disagree* to *strongly agree*. These issues includes broadband pricing, speed, quality of service (QoS), security, quality of perception (QoP), cultural quality, reliability, ease of installation, performance, and availability.

Research Findings

From a total of 25 participants, 20 responded to the questionnaires, of which all were used and analysed. A total of 12 (60 percent of sample size) male consumers completed and returned the questionnaire, whilst the female consumers were represented by a lower number of 8 (40 percent). Our analysis initially presents the profile of the respondents so as to give a picture of broadband consumers in the UK (Table 10.1). From the data gathered, most of the respondents (50 percent) belonged to the 35- 44 years age group; followed by the 24-34 years age group (30 percent). The age groups of under 16 years, 17-23 years, 45-55 years, and 56-70 had 1 respondent each (5 percent).

The income profile of the respondents spanned a wide range, with most of the sample (45 percent of respondents) being bunched in the medium income bracket (£10,000-39,999). At the other end of the spectrum the students sampled in our study held part-time jobs and indicated income levels of below £10,000, while white-collared professionals (doctors, IT consultants) and entrepreneurs indicated high income levels of income of above £70,000. Three of the 20 respondents confessed to still using dial-up to access the Internet at home, while the remaining 17 accessed the Internet via broadband connection. This shows that over 80 percent of respondents have taken up broadband. Thus, our study on their perspective on broadband quality is justified. Interestingly, an analysis of variance (ANOVA) run on the data has revealed that income levels do not affect broadband take-up; this, in our opinion, reflects the fact that broadband offering in the UK is competitively priced, and thus encourages a variety of users with different earning profiles.

Indeed, of the three people in our study who indicated that they were still using dial-up to access the Internet, two of them had earnings in excess of £50,000, while the third respondent earned less than £10,000. One of the respondents who used dial-up claimed not to know if he could get broadband in his home, while the two

Table 10.1. Profiles of respondents

Variable	Category	Frequency	Percentage
Borough	Respondents live in various London boroughs		
Sex	Male	12	60%
	Female	8	40%
Age	Under 16	1	5
	17 – 23	1	5
	24 – 34	6	30
	35 – 44	10	50
	45 – 55	1	5
	56 – 70	1	5
	Over 70	0	0
Occupation	Directors, doctors, lawyers, professors	3	15
	Electricians, mechanics, plumbers and other craft)		
	Managers, teachers, computer Programmers	7	35
	Machine operators, assembly, cleaning		
	Foremen, shop assistants, office workers	1	5
	Pensioners, casual workers		
	Unemployed		
	Student	6	30
	Others	2	10
Annual Income (£)	< 10,000	3	15
	10,000 – 19,999	3	15
	20,000 – 29,999	4	20
	30,000 – 39,999	2	10
	40,000 – 49,999	3	15
	50,000 – 59,999	2	10
	60,000 – 69,999	-	-
	70,000 – 79,999	1	5
	80,000 – 89,999	-	-
	90,000 – 99,999	1	5
	≥ 100,000	1	5

others indicated that they had access to the Internet in the office and that their need to access the Internet at home was minimal; it was thus more cost effective to access the Internet via dial-up. Indeed, all of these three respondents fell into the same *computer programmer, manager and teacher* category (see Table 10.1), supporting their claim that they access the Internet more in the office than at home.

Summary Statistics: Analysis of Level of Agreement Based on Perceived Broadband Quality

In our study, participants' responses were coded by awarding a numerical score of 1 to 7 to each of the expressed opinions on the Likert scale, with 1 representing an opinion of "strongly disagree" and 7 of "strongly agree." A t-test was applied to the survey data collected in order to establish whether the expressed opinions were statistically significant. However, we initially conducted a reliability analysis test on the data collected.

The reliability method used was the Alpha method. The Alpha is the most popular method of examining reliability (Hinton et al., 2004). The Alpha reliability method ranges from 0 for a completely unreliable test to 1 a completely reliable test (Hinton et al., 2004). Using SPSS we were able to analyse the data to show its reliability and the result is that, in the case of our survey, both the Alpha and standardized alpha are 0.8413 and 0.8528 respectively. This indicates that the responses received on the questionnaires sent out to the participants are reliable. Thus from this information, we can comfortably extract a fair perception on broadband quality from the respondents.

Services

Table 10.2 shows that the higher mean value of 6.15 and with a significant level of the 0.000, which indicates that participants were at the agreement end of the rating scale of 1 to 7. This signifies their intention to continue their broadband subscription. This comes as a result of the satisfaction of the broadband speed connection received from their ISP, which is reflected on the mean = 5.73 and p= 0.001.

Table 10.2. Indicators for ISP services

Quality factors	Indicators for Services (N)	Mean	Significant level (p)
Services: Satisfactory technical support from ISP	15	5.40	0.008
Services: Satisfied with the speed of broadband from ISP	15	5.73	0.001
Services: Not satisfied with the design of my current ISP web page	15	3.00	0.055
Services: It takes a long time to download from my broadband connection	15	2.47	0.007
Services: Satisfied with security measures from ISP	15	4.27	0.617
Services: Customer service representative handle my call quickly	15	4.87	0.072
Services: Customer service representative was knowledgeable	14	4.79	0.119
Services: Customer service representative was courteous	14	5.00	0.020
Services: Waiting for query answered is satisfying	14	4.57	0.302
Services: Automated phone system made customer service more satisfying	14	4.21	0.671
Services: Happy with overall customer satisfaction ISP	14	4.50	0.236
Services: Call was transfer to best person ale to answer request	14	4.64	0.108
Services: Happy with speed of response	14	4.79	0.127
Services: Intend to continue broadband subscription	13	6.15	0.000
Services: Considered moving subscription to another ISP	13	3.23	0.217
Services: Not happy with service provider	13	2.92	0.095
Services: Would recommend my service provider to my friends	13	5.62	0.001
Services: Broadband connection is reliable	14	5.57	0.002
Services: Overall service quality of current internet connection is satisfactory	13	5.08	0.042

Our results also highlighted that respondents would recommend their service provider to friends with participants indicating relatively strong agreement in this area (mean = 5.62; p = 0.001). This is as a result of the overall strong agreement of the respondents satisfaction on the service quality of their current broadband connection (mean = 5.08; p = 0.042) including a satisfactory technical support from their ISP (mean 5.40; p = 0.008). Most respondents strongly agree that the broadband connection is reliable (mean = 5.57; p = 0.002).

Usability

From our survey, most respondents disagree that the broadband connection is always breaking with a mean value of 2.21 and p = 0.002 (Table 10.3). This relates with the service section, which revealed that participants thought that their broadband connection was reliable.

Table 10.3. Indicators for usability

Quality factors	Mean	Significant level (p)
Indicators for Usability (N=14)		
Usability: Use broadband to listen to and download music	4.00	1.000
Usability: Enjoy using broadband to watch and download movies	4.00	0.151
Usability: Enjoy using broadband to play online games	3.00	0.089
Usability: Use broadband connection for videoconferencing	3.14	0.183
Usability: Satisfied with the quality of multimedia presentation	4.21	0.648
Usability: Able to understand the information content of multimedia display	5.00	0.068
Usability: Found broadband connection is always breaking	2.21	0.002

Table 10.4. Indicator for price

Quality factors	Mean	Significant level (p)
Indicators for Price (N-14)		
Price: Costly to subscribe to broadband at its current price	2.79	0.018
Price: Service provider offers good value for money	4.93	0.026
Price: Broadband price offered by service provider is expensive	3.21	0.102

Price

Table 10.4 reflects a strong agreement (mean =4.93; p= 0.026) of our respondents who indicated that the service provider offers good value for money which is also reflected in the disagreement that the current price they pay for their subscription are costly (mean =2.79 and p=0.018).

Government

Table 10.5 indicates the survey respondents' strong agreement to the fact that the UK government should promote broadband more (mean = 6.07; p = 0.000). This indication shows that it is important that the UK government promotes broadband more by providing additional funds towards broadband technology. In this respect, the UK government could follow the lead of their South Korean counterparts who promoted broadband by making funds available, which encourage the wide spread use of broadband connection in her country (Choudrie & Lee, 2004).

Concluding Discussion

In this chapter, we have presented the initial results of an ongoing research study that evaluates the broadband quality as perceived by users. The study highlighted that users' perceptions on quality do not vary too much from each other. This might be because broadband technology is still new in the market and that users are yet to be exposed to the full potential of broadband technology. This, however, could change when more data are gathered for the continuation of this research. However, we presented an integrated framework that would guide all stakeholders involved in broadband quality issues. This framework will be developed further in our future work.

Table 10.5. Indicator for government

Quality Factors	Mean	Significant level (P)
Indicators for Government (N =14)		
Government: UK government should promote broadband more	6.07	0.000
Government: Government should subsidise broadband	4.79	0.189

Despite the fact that broadband uptake has risen in the past years, our study has shown that there is still a lack of understanding amongst users and indeed service providers as to the potential and quality aspect of broadband as a whole. Moreover, our study also highlighted that users are only concerned about the basic broadband connection. This is because our research findings highlighted that most of the respondents used their broadband connection for email services followed by those who use their broadband connection for job search, educational research and listening to music. It was also discovered that most users are generally satisfied with the technical support service received from their service provider. This may be because most of the respondents are technically knowledgeable and are able to communicate issues encountered during their broadband connection to the support team of their ISP easier. However, they would not go for valued-added features offered by the service provider, which includes movies on demand, telephone calls as some of this tends to be highly priced by some service providers. This result, however, may be different if the survey is taken to a wider audience in our future work.

The next chapter (Chapter 11) will conclude the UK case study on broadband adoption, usage and impact. Chapter 11 will initially provide a summary, as well as conclusions drawn from each chapter of the UK case study. Following this, discussions on research contribution, limitations and further research directions are provided.

References

Chen, Y., & Lee, S.Y. (2000). Bounded tag fair queuing for broadband packet switching networks. *Computer Communications, 23*(1), 45-61.

Choudrie, J., & Lee, H. (2004). Broadband development in South Korea: Institutional and cultural factors. *European Journal of Information Systems, 13*(2), 103-114.

Choudrie, J., & Dwivedi, Y. K. (2004). *Broadband adoption: A UK residential consumers perspective.* ECIS, Americas Conference on Informations Systems.

Choudrie, J., Ghinea, G., & Weerakkody, V. (2005). Evaluating global e-government sites: A View using web diagnostic tools. *Electronic Journal of E-Government, 2*(2), 105-114.

Ghinea, G., Thomas, J.P & Fish, R.S. (1999). Multimedia, network protocol and users: Bridging the gap. In *Proceedings of the seventh ACM international conference on Multimedia International Multimedia Conference, (Part 1),* (pp. 473-476).

Hinton, P.R, Brownlow C., McMurray, I., & Cozens, B. (2004). *SPSS explained.* Rouledge: Taylor and Francis Group.

Howard, D. (2002). Reinventing broadband: Improved technologies, sober business plans, and tiered pricing may soothe an ailing industry. *netWorker, ACM press, 6*(2), 20-25.

Huber, P.W. (2002). *The Government's role in promoting the future of the Telecommunications Industry and Broadband Deployment.* Manhattan Institute for Policy research. Retrieved January 13, 2005, from http://www.manhattan-institute.org

Korakis, T., & Tassiulad, L. (2004). *Providing quality of service guarantees in wireless LANs complaint with 802.11e.* Computer Networks.

Krishnan, M.S. (1993). Cost, quality and user satisfaction of software products: An empirical analysis. IBM Centre for Advanced Studies Conference, In *Proceedings of the 1993 conference of the Centre for Advanced studies on collaborative research: Software engineering.* Vol. 1 (pp 400- 411).

Kumar. (2004). Broadband policy, Government of India, ministry of communication and IT, Department of Telecommunications.

Leung, K.K. (2002). Load-dependent service queues with application to congestion control in broadband networks. *Performance Evaluation, (Netherlands), 50*(1), 27-39.

Lorenz, D.H., & Orda, A. (2002). Optimal partition of QoS requirements on unicast paths and multicast trees.*IEEE/ACM Transactions on Networking (TON), 10*(1), 102-114.

Marchetti, C., Pernici, B., & Plebani, P. (2004). Quality of service: A quality model for multichannel adaptive information. In *Proceedings of the 13th International World Wide Web Conference on Alternate Track Papers & Posters* (pp. 48-4).

Orda, A. (1999). Routing with end-to-end QoS guarantees in broadband networks. *IEEE/ACM Transactions on Networking (TON), 7*(3), 365–374.

Pirsig, R.M. (1994). Zen and the art of motorcycle maintenance.

Romero, S. (2002). *Price is limiting demand for broadband.* New York Times. Retrieved from http://www.nytimes.com

Sangjo, O., Joongho, A., & Beomsoo, K. (2003). Adoption of broadband Internet in Korea: The role of experience in building attitudes. *Journal of Information Technology, 18*(4), 267-280.

Venkatesh, V., & Brown, S. (2001). A longitudinal investigation of personal computers in home: Adoption determinants and emerging challenges. *MIS quarterly, 25*(1), 71-102.

Chapter 11

Conclusion:
Contributions, Limitations, and Future Research Directions

Abstract

This chapter provides a conclusion of the results and discussions of the UK case study research presented in this book. The chapter begins with an overview of this research in the next section. This is followed by the main conclusions drawn from this research. Following this, a discussion of the research contributions and implications of this research in terms of the theory, policy and practice is provided. This is ensued by the research limitations, and a review of the future research directions in the area of broadband diffusion and adoption. Finally, a summary of the chapter is provided.

Research Overview

Chapter 1 defined the research problem and outlined the motivation for conducting this research. Given the large-scale investments in developing and upgrading broadband infrastructures, the slow uptake, and decreasing rate of residential broadband adoption in the UK provided the stimulus for this research. The literature analysis indicated that existing research on broadband had frequently examined this issue from macro level perspectives, such as the geography of the country, density of population, competition amongst Internet service providers (ISPs), local loop unbundling, stakeholder analysis, and government policies. Although these studies were helpful in the preparation and promotion of the broadband market, they offered limited implications to the ISPs who are the most important supply side stakeholders and who are directly involved with the household consumers.

However, the analysis of the literature suggested that an examination of broadband adoption, usage, and impact from the consumers' perspectives has just begun to emerge and is yet to be undertaken. It was also found that initial studies from the broadband consumers' perspective were mainly data driven and exploratory in nature. A recently conducted study (Oh et al., 2003) had made initial efforts in investigating the influence of the diffusion of innovation characteristics on attitude building towards the use of broadband in South Korea. However, the contributions of the study were also limited, as it did not examine how various factors including attitude influence the behavioural intentions (BI) and actual adoption of broadband. This study also excluded the issue of broadband usage and impact on consumers and how it differs from narrowband consumers. Since both these neglected issues are imperative for understanding broadband adoption, the implications of the study for ISPs and policy makers were limited.

Therefore, this research aimed to identify and determine the consumer level factors that influence broadband adoption and use and, consequently, its impact on UK household consumers. The objectives to achieve the overall aim included: developing a conceptual model; developing and validating a research instrument; conducting data collection and analysis in order to validate and refine the conceptual model; and finally outlining implications for theory, practice and policy. Chapter 1 also provided brief information on potential research approaches and the research contribution to theory, practice and policy.

In order to achieve the first objective of this research, Chapter 2 first reviewed the various technology adoption and diffusion related theories and models including the diffusion of innovations, TRA, TPB, DTPB, TAM, MATH, and use diffusion models. Since these adoption and diffusion theories and models provided this research with a number of underlying constructs or factors, they were considered to be guiding frameworks for this research. Chapter 2 then discussed and justified the reason for selecting factors that were expected to predict the BI to adopt broadband, which ultimately explains the broadband adoption behaviour (BAB). Also, the BAB is expected to differentiate between the rate and variety of Internet usage between broadband and narrowband users.

Using selected factors, a conceptual model of broadband adoption was developed. The proposed conceptual model is based on the assumption that the attitudinal, normative, and control factors are responsible for influencing BI to adopt broadband, which in turn is expected to predict the BAB. This study also included constructs that investigated whether broadband users differ from narrowband users with regards to usage and the impact of the Internet. Whilst discussing these factors, a number of underlying hypotheses were also proposed that were required to be tested in order to validate and refine the conceptual model. The accomplishment of Chapter 2 led to the achievement of the first objective of this research, which was *"to develop a conceptual model for examining consumer adoption, usage and impact of broadband."*

Chapter 3 provided an overview of the research approaches utilised within the information systems (IS) field and then selected an appropriate research approach for guiding this research. To validate and understand the conceptual model, it was found that a quantitative research approach would be more appropriate than a qualitative one. An overview of the underlying epistemologies was provided in order to decide whether positivism is appropriate as a philosophical foundation for this research. Following this, an overview discussion of various issues related to the available research approaches in the IS field, and a justification for the selection of the survey as a research approach, was provided.

The survey research approach was considered most appropriate when conducting this study. The survey approach facilitates data collection from a wide geographical area (i.e., nationwide) within a limited time with limited resources (Fowler, 2002). This was required to determine if the selected underlying constructs significantly explained consumers' decision regarding broadband adoption and subsequent use and impact. Following this, there was a detailed account of the various aspects of the survey approach. It was found that for the purpose of this research it was appropriate to employ UK-Info Disk V11 as a sampling frame, with stratified random sampling as a basis of sample selection and a postal (i.e., mail) questionnaire as a data collection tool. The reasons for these selections were also provided in a detailed manner. The issues relating to the data analysis were then discussed in detail and it was concluded that a number of statistical techniques such as factor analysis, t-test, ANOVA, χ^2 test, discriminant analysis, binary correlations, linear and logistics regression analysis were appropriate to utilise for data analysis purposes in this research.

Chapter 4 described the development process for the research instrument. The development process was achieved in three stages, which were made up of the exploratory survey, content validity, and instrument testing. The exploratory stage included surveying the known existing instruments, choosing appropriate items, creating the required new items and then determining if the selected items were appropriate enough to measure the perceptions of adopters and non-adopters. This stage also examined the reliability of the initial scale. At this stage it was found that although the majority of items either selected from the existing instrument or newly created ones were important enough to describe the behaviour of the adopters and non-adopters, the reliability of the scale was low in most cases. The exploratory stage also helped to identify a new construct that was subsequently termed as 'service quality.' The output of this stage of the research was utilised as an input for the content validation stage.

The content validation stage involved the creation of new items for each construct and then validation of their representativeness, utilising a quantitative approach. These new items were created utilising the items obtained from the exploratory stage and also re-surveying the literature and selecting the relevant items. In order to achieve the representativeness of the items, several experts working on broadband related issues evaluated the newly created items. This led to the calculation of a content validity ratio that was the basis of the exclusion or inclusion of the items. The outcome of this stage was the inclusion of representative items and the exclusion of non-related items.

The instrument testing stage was sub-divided into two stages, which included the pre- and pilot-tests. The purpose of the pre-test was to obtain feedback from the respondents on the instrument, in order to improve the wording of items. The purpose of the pilot test was to confirm the reliability of the items. The findings obtained from the pilot test demonstrated an acceptable level of reliability for all the constructs. The reliability of all the 10 scales improved after conducting a content validity and pre-test. This demonstrates the importance of performing a content validation for the increasing reliability of scale and also the representativeness of the items. The final output of the three-stage instrument development process that was undertaken in Chapter 4 is a parsimonious, 40-item instrument, consisting of 11 scales, all with a high level of reliability.

Chapters 5 and 6 presented the findings obtained from the data analysis of the conducted survey that examined consumer adoption, usage and impact of broadband in the UK households. The findings presented in Chapter 5 were obtained using several steps. The first step was to calculate the survey's response rate and conduct a response bias test. The estimated response rate was 22.4 percent and the response bias test suggested that there was no significant difference for the demographic characteristics such as the age and gender of the respondents and non-respondents. Also, the response bias test showed no significant difference between the responses of the respondents and non-respondents with regards to key constructs such as relative advantage, utilitarian outcomes, hedonic outcomes, service quality, secondary influence, knowledge, self-efficacy, and facilitating conditions resources. Responses for one of the construct 'primary influence' were found to be significantly different for the respondents and non-respondents. Since all other constructs and demographics were non-significant, it was concluded that there were minimal chances of a response bias in collected data.

Chapter 5 also presented the findings that illustrate the reliability test, construct validity and effect of ordering of the questionnaire items. The reliability test confirmed that the measures are internally consistent, as all the constructs possessed a Cronbach's alpha above 0.70. Construct validity was established utilising principal component analysis (PCA). Results of the PCA provided evidence of higher KMO values, a significant probability of Bartlett's test of sphericity, extraction of nine components consistent with the number of independent factors in the conceptual model (all nine factors possessed eigenvalues above 1), factors loaded above 0.40 and no cross loading above 0.40. This confirms that both types of the construct validity (i.e., convergent and discriminant) exist in the survey instrument. A t-test was conducted to confirm if any difference occurs due to the ordering of the questionnaire's questions. The results indicated that there was no significant difference between the responses with or without the questions being ordered. This further strengthened the existence of the construct validity in the survey instrument. The accomplishment of Chapters 3, 4, and 5 led to the achievement of the second objective of this research, which was *"to operationalise the constructs included in conceptual model by developing a research instrument and demonstrate their reliability and validity."*

Chapter 6 begins with the descriptive statistics for both the items and the scale. The findings suggested that the survey respondents showed strong agreement for all the constructs except the behavioural intention to change service providers (BISP). Results from the t-test and discriminant analysis suggested that there was a significant difference that occurred between the obtained responses from the narrowband and broadband consumers with regards to the attitudinal, normative and control constructs. Examination of the demographic differences that arose by employing the chi-square test suggested that the broadband consumers significantly differed from the narrowband consumers in terms of age, education, occupation and income. Finally, the regression analysis suggested that the attitudinal, normative, and control constructs significantly explained the BI that, in turn, significantly explained the BAB, along with the facilitating conditions resources.

The findings related to the usage of the Internet suggested that the broadband consumers significantly differed to the narrowband ones in terms of consumers' online habits and Internet use variations. The numbers of broadband consumers were exceedingly and significantly higher than the narrowband consumers for accessing or using 19 online services from a total of 41 services examined in this research. The last section of Chapter 6 examined the effects of broadband usage on a consumer's time allocation pattern in twenty daily life activities. The findings suggested that, for all twenty activities, the broadband consumers' time allocation pattern differed from the narrowband ones; however, a significant difference was found to be apparent for only five activities.

Chapter 7 presented the adoption and use of 41 electronic services and applications in the UK. The chapter concluded that the e-mail, search, storing, and displaying photos, downloading free software, listening to the radio, product purchasing, and using the Internet for fun emerged as the most commonly used online activities. Contrastingly, video conferencing, video streaming, online games, and video downloading are the least adopted online services amongst the UK's Internet users. Chapter 7 also concluded that traditional online activities such as e-mail, and search have become part of the daily life, while other activities such as online banking, share-market trading, education, and online music have become substantially popular. Alternatively, more technically advanced services such as video conferencing and streaming are in the initial stages of popularity amongst Internet users.

Chapter 8 discussed and reflected upon the findings from the theoretical perspectives.

Chapter 8 first presented and discussed the refined and validated conceptual model of broadband adoption. The discussion led to the conclusion that all the constructs except knowledge significantly explained the BI to adopt broadband, which in turn significantly explained BAB. Comparison of the adjusted R^2 obtained in this study compared with previous studies suggests that the performance of the conceptual model that can be used to understand the BI for the adoption and actual broadband adoption is as good as its guiding models. This chapter then discussed the usage and impacts of broadband on household consumers. The discussion revealed that the findings supported the assumption made in Chapter 2 that broadband users differ from narrowband users with regards to the usage and impact of the Internet. It is possible to conclude from the discussion that broadband consumers differ significantly from the narrowband ones in terms of duration and frequency of Internet access on a daily basis. It was also found that the use of broadband significantly affected the time allocation patterns on those activities that closely resemble it or those whose execution is dependent on faster access of Internet. Completion of Chapter 6, 7, and 8 led to achieving the third objective of this research, which was *"to empirically validate and refine the conceptual model that is proposed to examine broadband adoption, usage and impact in UK households."* Accomplishment of these three chapters also led to the answering of all the research questions outlined in Chapter 2.

Main Conclusions

The following main conclusions are drawn from this research and are based on the underlying research questions proposed in Chapter 2:

1. All three types of construct—overall attitudinal, normative and control constructs—significantly explained the BI of consumers when adopting broadband. Amongst these three types of construct, overall control construct contributed to the largest variance ($\beta = 0.367$) when explaining BI of broadband consumers. The overall attitudinal constructs contributed to the second largest variance ($\beta = 0.282$), whilst overall normative constructs ($\beta = 0.151$) contributed the least amongst the three types of constructs.

2. A total of seven constructs from attitudinal (relative advantage, utilitarian outcomes, hedonic outcomes), normative (primary influence), and control (knowledge, self-efficacy, and facilitating conditions resources) categories were expected to be correlated to the BI of consumers when adopting broadband in UK households. Of the seven constructs six, including relative advantage, utilitarian outcomes, hedonic outcomes, primary influence, self-efficacy, and facilitating conditions resources significantly correlated to the BI of consumers when adopting broadband in UK households. The only one that did not was knowledge.

3. In terms of the size of the effect of the six constructs that contributed significantly to the behavioural intentions, relative advantage exhibited the largest ($\beta = 0.255$) and hedonic outcomes ($\beta = 0.10$) demonstrated the least variance to the BI when adopting broadband in UK households. Primary influence explained the second largest variance ($\beta = 0.195$), which was followed by facilitating conditions resources ($\beta = 0.180$). The fourth strongest construct was self-efficacy ($\beta = 0.165$) and the fifth was utilitarian outcomes ($\beta = 0.113$).

4. Both BI and the control construct significantly correlated to the broadband adoption behaviour in UK households. In terms of the relative impact of these two constructs that contributed significantly to the BAB, BI had much higher impacts (**Exp (B)** = 2.50) than the control construct (**Exp (B)** = 1.57).

5. Amongst the five demographic factors that were included in this research, apart from gender, all the other four factors (i.e., age, income, occupation, and education) associated with the adoption of broadband in UK households.

6. Although both service quality and secondary influence constructs were significantly correlated to the BI when changing current service provider (BISP), the overall explained variance was low (adjusted R square = 0.074). This meant that both these constructs were also not able to explain the variation of BISP.

7. The rate of Internet usage differs for the broadband and narrowband users. Broadband users spend more time and access the Internet more frequently than narrowband ones.

8. Broadband users access significantly more online activities than narrowband users.

9. The use of broadband is associated with the time spent on various daily life activities. However, the magnitude of this association differs according to the nature of the activities. The association was more on activities such as an increase in working at home, a decrease in shopping in a physical store and a decrease in working in the office compared to activities, such as spending time with family and friends.

Research Offerings and Implications

This research presents one of the initial efforts towards understanding the adoption and usage behaviour and impact of broadband in UK households. Furthermore, this study is one of only a few studies that address the issue of individual adoption and usage of technology in the household that is beyond the boundary of workplace. The following statement illustrates the importance of conducting a study that examines technology adoption in the household context:

Much work in the area of home computing remains to be done to better understand the adoption of technologies at home given its potential implications for society in general and for the workplace in particular. (Venkatesh & Brown, 2001)

By employing the quantitative approach, this study provides an initial effort that confirms the role of various constructs (such as relative advantage, utilitarian outcomes, hedonic outcomes, primary influence, secondary influence, knowledge, self-efficacy, and facilitating conditions resources) for understanding the adoption of broadband within households. This study not only initially validated the model of adoption of technology in households (MATH) for examining broadband, but also extended it by incorporating constructs such as relative advantage, service quality and also the decomposition of social influence into two categories, namely primary influence and secondary influence.

The findings of this study empirically suggest that primary influence plays a key role in the first time adoption decision of broadband in households, whilst secondary influence emphasises a key role of continued subscription with the same service providers. An added strength of this research is that, unlike other related studies on technology adoption, the current work investigated all three components of diffusion comprising adoption, usage and impact. This in turn helped in obtaining an extensive understanding of adoption and usage from the perspective of household consumers. This study brings about several theoretical contributions and implications to practice and policy.

Offerings to Theory

The first offering of this research towards theory is that it integrates the appropriate information systems (IS) literature in order to enhance the knowledge of technology adoption from the consumer perspective. That is, initially, this research evaluates the flexibility of various models when studying the technology adoption issues. Second, it assimilates previous research findings to develop a coherent and comprehensive picture of the technology adoption research conducted in the IS field. Third, this research introduces a conceptual model that integrates factors from different technology adoption models in order to study home technology diffusion from a consumer's perspective.

The second offering is to empirically confirm the appropriateness of various constructs and validate the conceptual model in the context of household consumers. Venkatesh and Brown's (2001) study was based on the semi-structured interview method for data collection, and examined PC adoption within American households. The study advised researchers to move beyond and employ a survey research approach for validating the conceptual model in order to utilise it for examining technology adoption from a household consumer's perspectives. This is emphasised in the following future research direction of the study:

A study that employs a survey with items/questions measuring the various dimensions/constructs will help achieve methodological triangulation and examine generalisability. Such a quantitative study will also help shed more light on the asymmetry between the belief structure influencing intenders and non intenders. (Venkatesh & Brown, 2001)

This study employs a survey with items/questions measuring the various constructs such as relative advantage, utilitarian outcomes, hedonic outcomes, primary influence, secondary influence, knowledge, self-efficacy, and facilitating conditions resources specifically in the context of broadband adoption. Therefore, the approach of this study is similar to that suggested by Venkatesh and Brown (2001). This quantitative study clearly illustrated the asymmetry between the factors influencing the adopters (i.e., broadband consumers), and non-adopters (i.e., narrowband consumers).

The third offering is that this research introduced and validated novel constructs such as service quality, secondary influence, and perceived behavioural intention to change service provider for investigating the continued adoption of broadband in households. Since these constructs were not included in any of the guiding frameworks, including the TPB, DTPB, MATH, and Diffusion of Innovations, there is an offering that contributes towards theory development in the form of theory extension. These constructs can be utilised to measure the perception of household consumers towards the continued adoption of subscription-based emerging technology and new Internet based services.

The fourth theoretical offering of this research was to confirm the role of socio-economic variables such as age, education, income, and occupation in explaining the actual adoption of broadband. This was considered to be one of the important research issues of this study, which is to investigate the socio-economic characteristics in order to determine their effect on the adoption of broadband. This study concludes that age, education, income and occupation are important variables that distinguish broadband's adopters and non-adopters, although gender does not. Referring to the research question that was posed initially in Chapter 2, it was learnt that there is the occurrence of an unequal or heterogeneous adoption rate or digital divide in various variables, including age, income, occupation, and education.

The fifth contribution of this research towards theory is to successfully utilise theoretical constructs such as the rate and variety of Internet use to examine the differences between narrowband and broadband consumers. Previous studies that focused upon the usage of broadband in households contained two main limitations. First, they were data driven and exploratory in nature, therefore they lacked a theoretical underpinning. Second, these studies either examined broadband or narrowband consumers, therefore they lacked a cross-sectional approach to distinguish the broadband consumers from narrowband consumers.

This research adapted the usage constructs from the use diffusion model to distinguish in terms of usage the broadband consumers from narrowband consumers. Therefore, by overcoming the two abovementioned limitations of previous studies, this research provides better understanding of Internet usage and helps to enhance the theoretical foundation.

The sixth theoretical contribution of this research is to develop theoretical understanding of the impact of broadband use in households. The discussion on the impact of broadband in Chapter 8 suggested that the findings supported the overall assumptions made, which were illustrated in Figure 8.4, and the concept of homeostasis in household systems (Robinson, 1977; Vitalari et al., 1985). Therefore this study provides an incremental contribution towards theory development in the area of the impact of ICTs in households. This study also contributes to academia by confirming the findings of previous studies relating to the impact of broadband related issues and also by analysing the differences in the behaviour of broadband and dial up users.

The seventh offering towards theory is the analysis of the approaches that are utilised to study technology adoption and diffusion related issues. Several such studies were conducted to review research approaches employed within the IS area and were considered important, as they provided further direction towards the utilisation of available research approaches. However, no such effort had been made to review research approaches specifically employed to examine technology adoption and diffusion related issues. The current study conducted a review of articles related with technology adoption and diffusion published in four respected and peer reviewed journals including *MIS Quarterly, European Journal of Information Systems, Information Systems Research* and *Information Systems Journal*. Details about this process and findings were presented in the 'Research Methodology' chapter (Chapter 3). By doing so, this study contributed by reflecting upon the use of various research approaches for investigating technology adoption and diffusion issues from the perspectives of both users (i.e., in organisations) and consumers (i.e., in households).

The eighth theoretical offering is the development and validation of a survey instrument. In a situation where theory is advanced, but prior instrumentation is not developed and validated, it is essential to involve the creation and validation of new measures and such efforts are considered to be a major contribution to scientific practice in the IS field. As Straub et al. (2004) stated:

Researchers who are able to engage in the extra effort to create and validate instrumentation for established theoretical constructs are testing the robustness of the constructs and theoretical links to method/measurement change. This practice, thus, represents a major contribution to scientific practice in the field. (Straub et al., 2004)

Moore and Benbasat (1991) and Davis (1989) provide examples of such work on instrument development and validation, which the authors considered as a major contribution towards the IS field. Although the constructs utilised in this research have been taken from established theories and models such as the TPB, DTPB, MATH, and the diffusion of innovations, prior instrumentation to study broadband adoption and diffusion was not developed and validated in existing studies. Therefore, it was considered essential to create and validate a new research instrument for constructs included in the conceptual model.

Since this study satisfied the criteria quoted above (Straub et al., 2004), it makes a substantial contribution towards the research methodology. This was achieved by modifying, creating and validating measures that correspond to various constructs included in the conceptual model. The research instrument developed and validated in this research can be utilised to examine various emerging technologies within the context of households.

Offerings to Industry and Policy

This study concludes that the amongst seven independent variables, six constructs including relative advantage, utilitarian outcomes, hedonic outcomes, primary influence, self-efficacy, and facilitating conditions resources significantly influence the BI of consumers when adopting broadband in UK households. The only exception is knowledge. It was found that relative advantage was most important and hedonic outcomes least important in terms influencing the BI when adopting broadband in UK households. Other important constructs that fall within two extremes were primary influence, facilitating conditions resources, self-efficacy, and utilitarian outcomes. The findings of this research generate a number of issues that may assist both policy makers and ISPs for increasing consumer adoption of broadband.

For example, since relative advantage is found to be the strongest construct, it indicates that ISPs have to provide broadband services to consumers in such a package that would illustrate a clear advantage over narrowband consumers. Similarly, facilitating conditions resources is the third most important factor in terms of influencing BI to adopt broadband. This has implications for both ISPs and policy makers. For instance, ISPs have to think about more consumer centric services and alternative price plans so that all consumers who want to subscribe broadband would be able to do so. Policy makers have to provide alternative places for broadband access where lower income groups or those who cannot afford it can use high speed Internet. It may help to increase behavioural intention to adopt broadband and therefore encourage overall adoption and diffusion of broadband in the UK household.

As previously mentioned, self-efficacy is also an important factor that influences behavioural intention to adopt broadband, which brings policy-related issues. This suggests that there is a need to equip citizens with the skills to use computers and the Internet. Since both utilitarian outcomes and hedonic outcomes are important factors for explaining behavioural intentions, it is important to integrate more content and applications for the purpose of household and entertainment utility.

One of the research questions of this study was to investigate the socio-economic characteristics in order to determine their impact on the adoption of broadband. Referring to the research question that was posed initially in Chapter 2, it was identified that there is the occurrence of an unequal or heterogeneous adoption rate or digital divide in various dimensions including age, income, occupation, and education.

It has been learnt that an important implication for supply side stakeholders, such as policy makers and industry, is to identify segments of society that are slow in adopting broadband. By obtaining results such as those afforded by this research, the reasons for slow adoption can be explored and appropriate measures can be developed and implemented so that they can be overcome.

As discussed before, broadband service providers may face two key challenges. First, there are consumers who cannot afford the current price plan. Therefore, the ISPs may consider providing alternative price plans in order to create a mass-market demand, which is an issue currently being emphasised. Second, some of the consumers with a high annual household income may also be reluctant to subscribe to broadband due to a lack of compelling content; hence, the challenges to the ISPs are to integrate content and applications and make them apparent to the ordinary members of the public.

The ISPs may overcome these challenges by offering differential price plans and segment specific broadband subscription packages. For example, the ISPs can differentiate within the offered subscription price ranges depending upon factors such as the income levels and needs of the users. Further ideas of fulfilling this challenge are as follows. First, with an increasing demand in the lower income segments and those with fewer needs of broadband, ISPs may offer price plans that can compete with the current price plan of un-metered narrowband.

Currently, there is a price gap between the two packages; therefore, a low price plan of un-metered narrowband is an inhibiting factor for broadband adoption in the segments with lower incomes and fewer needs. Second, since cost is not a factor of consideration when segmenting between the higher income and occupation levels, it should be offered in broadband packages with even faster speeds and appealing content. Such packages may assist in illustrating the clear benefits of broadband over narrowband to consumers of higher incomes and occupation levels, and provide them with added reasons for subscribing to broadband.

A recently published report that was submitted to the Department of Trade and Industry (DTI) suggests that the penetration of broadband is likely to promote usage of advanced Internet content and applications. However, due to the lack of data at present, it is difficult to support this theoretical claim (Analysys, 2005). The Analysys report states that:

Much has been made of the requirement for countries to invest in broadband communications infrastructure, and to promote its usage. Increased take-up of broadband access services is expected to stimulate usage of advanced Internet content and applications by consumers and by businesses, thus changing individuals' behaviour, creating new industries, or increasing productivity in existing industries. However, data to prove the theory is hard to come by. (Analysys, 2005)

The Analysys (2005) report supported the above claim on the basis of the secondary data obtained from the studies conducted in the USA. This research is one of initial efforts that provides primary data in the context of the UK and supports the aforementioned theory. The findings of the current study clearly illustrate that the UK broadband consumers do tend to use the Internet differently to the narrowband consumers. Both the rate of and variety of Internet usage is significantly higher for the broadband consumers in comparison to the narrowband ones. This means that broadband consumers use the Internet more than the narrowband consumers.

Furthermore, this study also examined the usage of 41 online services and applications. The study suggests that for all online services the numbers of broadband consumers were higher than the total number of narrowband ones. It was also illustrated that for the usage of many services, the differences between the numbers of broadband and narrowband consumers were significant. Thus, the findings of this study may assist in justifying investment in the area of broadband deployment and assist policy-making organisations such as DTI and Ofcom that are involved in the development and deployment of broadband in the UK.

The findings also have important implications for the electronic mass media and telecommunication industries. The mass media industry is likely to benefit from the diffusion of broadband, as more respondents use online material rather than utilising traditional reading resources. This may encourage the online media industry to attract revenues from advertising and subscription fees. According to the findings, receiving and making phone calls have decreased for both narrowband and broadband consumers.

Therefore, the telecommunication industry may have to transform in terms of its business model. For example, the new business model may find a means of pricing consumers at instances of phone calls, which makes using VoIP or other emerging applications a consequence of a broadband environment.

Bearing these points in mind, it can be argued that the contributions of this research are substantial for both policy makers and the telecommunication industry. Therefore, this research is viewed to be pertinent for the current period of broadband deployment. The discussion on contributions and implications of this research has led to achieve the fourth objective of this study, which was "*to provide implications for practice and policy that may encourage consumer adoption and use of broadband.*"

Research Limitations

One of the limitations of this study was related to the availability of the sample frame. The electoral register is considered to be the most comprehensive sample frame for the UK population. However, it could not be employed to obtain the respondent addresses due to recent legal restrictions that have been placed upon utilising the electoral register for any data collection purposes. The UK-Info Disk V11 database, which was utilised as a sample frame for this research, is as comprehensive as the electoral register. This is because it is derived from the electoral register and is regularly updated. However, it does not provide addresses in the form of lists; instead it requires an initial extraction of addresses, creation of a list and then utilisation of that list. Since the UK-Info Disk V11 does not allow extraction of all addresses available in the database, it excludes some parts of the UK population from being included in the sample frame. This limited the ability to generalise the findings for the entire UK population. Hence, even though responses were received from various parts of the UK, the findings of this study were considered to be a representative sample from which respondents were selected, instead of the whole UK population.

Although the response rate that was obtained in this research was considered acceptable in IS research, there was the possibility of a non-response bias. Although the possibility of a non-response bias cannot be ruled out completely, the non-response bias test undertaken in this research suggests that it had minimal chances of affecting the findings.

This study provides a snapshot of broadband adoption, usage, and the impact of broadband in the UK households. The findings may change as technology becomes established and consumers become more experienced in its use. However, as this research has a limited completion timeframe, it is not possible to conduct further data collection in order to observe the effect of time on the adoption, usage and impact of broadband.

This study was focused upon utilising a quantitative approach that may have limited the ability of this research when attempting to obtain an in-depth view of household technology adoption and usage. However, due to the time and resources there was a constraint such that it was not possible to conduct both qualitative and quantitative research.

The data for the current study was collected using a postal questionnaire, which limited the ability to include important variables such as family life stage/life cycle. This variable considers family as a consuming unit that progresses through a series of stages and is characterised by disposable income and the total number of members in a family. As this variable is an incremental one, it offers a deeper and richer understanding. Therefore, this variable will be very useful to investigate why a particular age or income group is more oriented towards adopting broadband than others. Additionally, it will be useful to collect in-depth data by conducting interviews that examine variables such as the life stage and relationships amongst age, education, income, and social class. This will provide a clearer and more complete picture of broadband adoption and will certainly be helpful in developing a further understanding of the critical segments for subject areas related to diffusion and marketing.

The impact component of this study was constrained by the following limitations. First, the findings were generalised to the household. The data was collected from only the household head in order to examine the impact of broadband upon the online habits and time allocation patterns. However, the behaviour of the household head may differ to the other members of the family. Second, the study is based upon self-reports and recall measures rather than observation or diaries. Finally, although the study showed the impact of broadband, it did not reveal why and how that occurs; therefore, this is difficult to explore using only a survey research approach.

Future Research Directions

 With regards to adoption and usage in the future, this research intends to examine whether the findings obtained from this study are specific to UK households or whether the results will be the same across other countries. This would require a cross-cultural approach when understanding broadband adoption. This, to a certain extent, has been covered by a Dutch case study presented in 'Division II' and the developing country case study presented in 'Division III' of this book. However, further efforts require examining such issues across the countries.

The questionnaire findings would have been strengthened if it had been possible to also supplement them using interviews. As mentioned in the previous limitations section, this supporting tool had to be abandoned due to the limitations of time and resources. The findings would also have been reinforced if the research had been a longitudinal one. The data for this research has been collected over a short period of time and provides a snapshot. However, it could be expanded over a longer period of time to offer a longitudinal study. Further justification for undertaking a longitudinal study is the reasoning that the elimination of any variables could achieve anomalies in the obtained results.

The acknowledged limitations to study the impact of broadband can be overcome by conducting a longitudinal and qualitative enquiry that will employ a combination of data collection tools such as diary, observations, interviews and questionnaires. The research suggests a longitudinal, qualitative study to be an appropriate future direction to overcome this limitation. This will allow an in-depth understanding of the impact of broadband and its implications in the context of the household to be obtained. Further, this will lead to an examination of the usage and impact of broadband differing amongst the various members of a household.

Due to the emergence of business to consumer (B-2-C), consumer to consumer (C-2-C) electronic commerce and e-government services, there is now an emphasis upon the diffusion of high speed Internet; therefore, studying the impact of broadband on household consumers becomes a very broad area. There is a need to research specific areas such as new communication methods, music and software downloads, entertainment, retail, travel, and tourism on an individual basis in order to determine the real impact of broadband.

Furthermore, there is a need to explore associated issues such as the positive and negative impact of these changes on the growth and development of the Internet, the diffusion and sustainability of broadband technology, family life and work, social interaction and development, and growth of the business-to-consumer, consumer-to-consumer electronic commerce and e-government services areas.

Finally, this research was focused upon considering the advantages of broadband due to the slow rate of adoption and its effect on adoption and diffusion of new electronic services. There are several negative aspects of broadband adoption; however, this issue was not included within this study due to time and resource constraints. An example of the negative aspects of broadband adoption includes the problem of online security in terms of transactions such as online shopping and online banking. Furthermore, broadband also raises the issue of the security of children accessing the Internet at home (Udo, 2001). Therefore, it is advisable that future research may take issues such as privacy and security into consideration when examining the adoption and usage of broadband.

Summary

This chapter provided an overview and conclusion to the results and discussions of the research presented in the first division 'the UK case study' of the book. First, the contents of each chapter were discussed briefly followed by a drawing together of the main conclusions of this research. This was followed by a discussion of the research contributions and implications that this research has made in terms of theory, policy, and practice. Following this, the research limitations were listed. Finally, the future research directions in the area of broadband diffusion and adoption were provided. The two chapters of the next division (Division II) will examine and discuss the issues pertaining to broadband adoption, usage and impacts in the Netherlands.

References

Analysys (2005). Sophisticated broadband services. Retrieved July 15, 2005, from *Final Report for the Department of Trade and Industry* at http://www.egov-monitor.com/reports/rep11610.pdf

Davis, F. D. (1989). Perceived usefulness, perceived ease of use, and user acceptance of information technology. *MIS Quarterly, 13*, 319-340.

Fowler, F. J. Jr. (2002). *Survey research methods.* London: Sage Publications.

Moore, G. C., & Benbasat, I. (1991). Development of an instrument to measure the perceptions of adopting an information technology innovation. *Information Systems Research, 2*(3), 192-222.

Oh, S., Ahn, J., & Kim, B. (2003). Adoption of broadband Internet in Korea: The role of experience in building attitude. *Journal of Information Technology, 18*(4), 267-280.

Robinson, J. P. (1977). *How Americans use time.* New York: Praeger.

Straub, D. W., Boudreau, M-C., & Gefen, D. (2004). Validation guidelines for IS positivist research. *Communications of the Association for Information Systems, 13*, 380-427.

Udo, G. J. (2001). Privacy and security concerns as a major barrier of e-commerce: A survey study. *Information Management and Computer Security, 9*(4), 165-174.

UK-Info Disk V11. Retrieved from http://www.192.com/products.cfm?icdaction= details&item_id=41.

Venkatesh, V., & Brown, S. (2001). A longitudinal investigation of personal computers in homes: Adoption determinants and emerging challenges. *MIS Quarterly, 25*(1), 71-102.

Vitalari, N. P., Venkatesh, A., & Gronhaug, K. (1985). Computing in the home: shifts in the time allocation patterns of households. *Communications of the ACM, 28*(5), 512-522.

Division 2

The Dutch Case Study

Chapter 12

A Longitudinal Study to Investigate Consumer/User Adoption and Use of Broadband in the Netherlands

Karianne Vermaas, Dialogic/Utrecht University, The Netherlands

Lidwien van de Wijngaert, Utrecht University, The Netherlands

Abstract

The Netherlands has experienced a high uptake of broadband Internet amongst its population. However, the question remains whether there are any differences between people with a broadband connection to people with a narrowband connection. The central research question in this chapter is therefore: how do Dutch internet users with a broadband connection differ from people with a narrowband connection in terms of demographics (age, gender, education), internet experience (experience, frequency, intensity of use), expectations (of narrowband users), experiences (of broadband users), annoyances, and patterns of internet usage? Secondly, this chapter addresses the question of whether and how these differences change over time. The chapter uses a model of technology adoption and use that is built upon different theories such as diffusion of innovations, uses and gratifications, and media choice theory. The results are based on two online questionnaires, in 2003 and 2004/2005 (N = 2404 and N = 1102) with regard to current Internet behaviour in the Netherlands. Results show that broadband users are heavier Internet users and that broadband technology is mostly a matter of comfort, not of complete new ways of using the Internet.

Introduction

In December, 2005, the Netherlands was indicated as one of the four leading countries with regard to broadband penetration, with more than 25 subscribers per 100 inhabitants (OECD Broadband Statistics, December 2005). That is why the Netherlands form a good starting point for investigating the usage of broadband and how it differs from the use of narrowband Internet. Other (EU) countries can anticipate future developments based on the Dutch experiences.

Broadband has potentially many advantages for Internet users. The particular aspects of broadband, such as always-on and the ability to send and receive large amounts of data, may provide the user with more convenience compared to people who have used traditional telephone lines. Broadband users could save a considerable amount of time (Pociask, 2002) compared to narrowband users. According to Anderson et al., (2002), broadband users make more frequent use of a wider range of applications. In research conducted by Wales, Sacks and Firth (2003), respondents universally said that they were not driven to adopt broadband by a specific application. Rather, they found that broadband enabled them to use standard Internet applications (email, chat, browsing) more efficiently. Wales at al., (2003) speak of killer attributes instead of killer applications and these include reliability of service, speed of downloads, networking of home computers, and the convenience of always-on connectivity.

Contrary to Wales et al. (2003), there are researchers that do consider specific broadband applications, including games, swapping of large files, and pornography to be the killer applications for broadband to the home (Anderson et al., 2003; Firth & Kelly, 2001; Thierer, 2002; Wales, 2002). Another possible reason for people to switch from a narrowband connection to broadband may be due to its flat fee. Especially for heavy users, a flat fee may help to reduce their internet costs (Wales et al., 2003). The study of Wales et al. (2003) also observed some discrepancies between the perceived benefits in households that had adopted broadband and those that had not. Furthermore, there are researchers that suggest that Internet and particularly broadband enhances the quality of life by increasing contact between parties. Internet users in fact have more face to face and phone contact with friends and family than do non-users. Those with good conventional friendship links extend those to the

Internet (Katz et al., 2001). Also, online relationships have been found to enhance conventional relationships (Katz & Rice, 2002; Wellman, 2001).

Broadband Defined by its Characteristics

In this research the term narrowband is used for any connection that is established through dial-up access (traditional modems and an analogue telephone line). Therefore public switched telephone network (PSTN), and also integrated services digital network (ISDN) connections are considered as narrowband. Two important characteristics of broadband Internet are 'flat fee' and 'always-on.' 'Flat fee' means that the subscribers pay a fixed amount of money per month, regardless of the actual time spent online (as opposed to dial-up access, where people pay per time-unit that they are connected). 'Always-on' means that there is direct connection at any time; there is no need to dial up. Based on these characteristics asymmetric digital subscriber line (ADSL), cable, and fibre optics are considered as broadband in this research.

The User Perspective

Despite the fact that many service providers, network operators, and researchers claim to maintain a user perspective, much of the research and service development is still inherently technology driven. In many cases, technology is developed using an iterative process in which technology is presented to users (in early and/or later stages) who can state what they like and dislike about the technology. This method of technology development does not address the question of whether people actually need that technology. So, instead of looking at what technology can do for people, we look at user behaviour, wishes, and frustrations in a daily context in order to find out what technology people may need. Based on a broad set of theoretical insights we will try to explain the use of broadband technology.

The following section provides a brief discussion on the theoretical framework. After that the research methodology is discussed, followed by the findings. Finally, a concluding discussion of this chapter is provided.

A Basic Model of Technology Adoption and Use

In our opinion, technology adoption and use is a process that results from experience and incentives from the past and present. In several research areas theories with a similar starting point can be found. In organizational literature innovation is described in terms of phases, starting with adoption (the strategic decision to invest in new technologies), implementation (putting everything in place) and incorporation (usage on a daily basis) (Andriessen, 1994). In the area of communication theory Trevino, Webster and Stein (2000) distinguish between general media (i.e., both old and new (information) technology) use and media choice. General use refers to an individual's broad pattern of technology usage over time. Choice refers to an individual's specific decision to use a technology in a particular communication incident. When we combine these concepts the model as depicted in Figure 12.1 can be drawn.

The general outline of the model is as follows. Adoption is the process in which an individual comes to the decision to start using a new technology. Adoption is related to a relatively slow, long term decision making process, learning to use and adopt that new technology. Major influences are demographic characteristics, attitudes, and self-efficacy. As a result of adoption an individual starts using that technology. Use is the case by case decision whether or not to use the adopted technology or some alternative. Major influences here are task characteristics, contextual determinants and social influence. As a result of the use, future use can be influenced. Because a user gets more experienced, usage will develop and change. Use (and the experience that is the result) of technology will also influence the adoption of other (new) technologies in the long run.

The model that we use here is basically a synergy of many other theories (Figure 12.2, including main references). Diffusion of innovations focuses strongly on the adoption phase. The technology acceptance model (TAM) also focuses relatively strongly on the adoption phase. The difference between the two theories is that diffusion of innovations describes this from a general perspective whereas TAM has the individual user as a starting point. Domestication describes how technology use changes over time and is placed on a middle position for both dimensions. Uses and gratifications provide a general framework to think about the relation between needs and technology use. Media choice theory fills this in by relating (perceived) characteristics of the task to characteristics of the technology.

Based on this overview it is possible to identify several variables that explain the differences between the adoption and use of narrowband and broadband technol-

Figure 12.1. A basic model of technology adoption and use

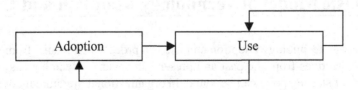

Figure 12.2. A two-dimensional space of adoption and use theory

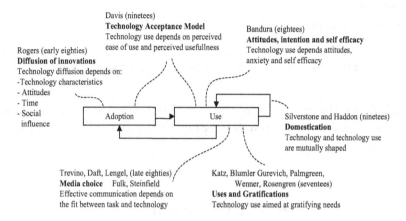

Table 12.1. Relation between theories and research variables

Focus of theory on variables is ◎ Strong ⊙ Moderate ○ Light - None	Demographics (e.g. age, gender, education)	Internet experience (e.g. internet experience, frequency, intensity of use)	Broadband experience (e.g. expectations, experiences and irritations)	Patterns of internet use (e.g. task characteristics)
Diffusion of innovations	◎	⊙	⊙	-
Uses and Gratifications	⊙	⊙	⊙	◎
Media choice theory	⊙	⊙	⊙	◎
Technology Acceptance Model	○	◎	⊙	◎
Attitudes and self efficacy	○	⊙	◎	-
Domestication	○	⊙	◎	-

ogy. In Table 12.1, the relation between the theories that were presented and the variables that are chosen is depicted.

Based on these findings, the following research questions have been derived.

How do users with a broadband connection differ from users with a narrowband connection in terms of:

• Demographics (age, gender, education)
• Internet experience (experience, frequency, intensity of use)

- Expectations (of narrowband users) and experiences (of broadband users) and irritations?
- Patterns of internet usage?

And (how) did these differences change over the period 2003 until 2005?

The goal of answering these questions is to obtain further insight into the theoretical framework that was presented as well as the future of adoption and use of broadband technologies.

Research Methods

Longitudinal

The data for this chapter is collected in a longitudinal study that allows us to see how technology use is developing over time. The first wave of data gathering used for this chapter took place from January to March, 2003. The response consisted of 2,404 completed and usable questionnaires. The last measurement was from October, 2004, to February, 2005, and resulted in 1,102 completed questionnaires.

Multi-Method

We used a multi-method approach, resulting in the following research lines. The first research line was an online questionnaire. The objective of this survey is to obtain insight into current internet behaviour. Questions regard type of internet access, activities on the internet (information seeking, communication, entertainment, and transactions), skills and experiences, wishes, expectations, and the reasons for and impediments to switching to a broadband connection. In order to obtain a broader, more in depth view of how technology is used we also conducted two other research lines: a diary project and broadband focus groups. The focus of this chapter lies with the results of the online questionnaire.

In 2003, 25 percent of the respondents had a narrowband connection and 75 percent of the respondents a broadband connection. In 2004-05, 16 percent had a narrowband connection and 84 a broadband connection.

Results

In this section we will focus on the differences between users with a narrowband connection and users with a broadband connection to the Internet. We will describe the differences in demographics, experience (including irritations, expectations, and perceived barriers and benefits of broadband) and finally we will focus on usage patterns.

Demographic Differences

Both narrowband and broadband users are more often men than women in this sample. The most important difference between narrowband and broadband is that

Table 12.2. Differences in demographic characteristics between narrowband and broadband

Demographic characteristics 2003		Narrowband	Broad-band	Statistic	Sign. 2-tailed
Age		$\mu = 40.8$	$\mu = 40.3$	$t = 0.74$	0.46
Income		$\mu = 4.3$	$\mu = 4.3$	$t = -0.09$	0.93
Education		$\mu = 3.2$	$\mu = 3.1$	$t = 2.22$	**0.03**
Gender (%)	male	73.1	83.7	$chi^2 = 33.18$	**0.00**
	female	26.9	16.3		
Household (%)	one person	28.5	24.9	$chi^2 = 9.21$	**0.03**
	single parent	4.0	3.6		
(Un)married couple. No children		31.2	28.3		
(Un)married couple. With children		36.3	43.3		
Demographic characteristics 2004/5		Narrowband	Broad-band	Statistic	Sign. 2-tailed
Age		$\mu = 45.6$	$\mu = 42.8$	$t = 0.84$	0.18
Income		$\mu = 6.3$	$\mu = 5.6$	$t = -1.98$	0.88
Education		$\mu = 3.0$	$\mu = 3.1$	$t = 2.15$	**0.03**
Gender (%)	male	68.3	78.7	$chi^2 = 28.12$	**0.00**
	female	31.7	21.3		
Household (%)	one person	28.7	19.8	$chi^2 = 5.43$	0.12
	single parent	1.2	3.7		
(Un)married couple. No children		37.8	34.7		
(Un)married couple. With children		32.3	41.6		

this image is even stronger for broadband users. There are no significant differences in age and income between narrowband and broadband users. In 2003, the education level of narrowband users was higher than that of broadband users. In 2004-05, this was the opposite. The overall impression is that broadband users (in the Netherlands) do not match up to the image that is often depicted of broadband users: they are not per se better educated, younger and with higher incomes. Finally, among narrowband users there are more one person households and households without children. The highest penetration of broadband connections can be found within families with children.

Differences in Experience

The image of broadband users in 2003 is quite similar to the image of broadband users in 2005. Broadband users are significantly heavier users of the internet. They perceive themselves as more experienced users and use the internet more often and have longer online sessions than respondents with a narrowband connection.

If we compare the two years, we find most changes among narrowband users. In 2004-05, narrowband users are more often online once a day (this was in 2003 with 20 percent, whereas in 2004-05 there were almost 41 percent). Also, in 2004, among narrowband users there are relatively more new and less experienced users. A possible explanation is that many narrowband users that were already quite experienced Internet users (both in years as in their description of their internet capabilities) in 2003 have adopted a broadband connection in the following year. In 2004, only the new and less experienced Internet users stuck to their narrowband connection. In Table 12.3 these results are summarised.

Annoyances

Annoyances could be an explaining factor if we look at the reasons why people adopt a faster connection with more bandwidth. This is why, before asking about explicit reasons and thresholds, we have asked the respondents to state the biggest annoyances they have with their current connection. We see (Table 12.4) that narrowband users, especially in 2003, are annoyed by a slow connection and high costs. However, the discrepancy between the annoyances in cost of broadband and narrowband users is getting smaller over the years. If the group of narrowband users, annoyed by speed and high costs feels that broadband can change that, they will probably be willing to make the step to broadband. On the other hand broadband users suffer more from technical failures, lack of privacy and spam. This possibly could be reasons for narrowband users to not want to make the step to broadband.

Table 12.3. Difference in Internet use between narrowband and broadband

	2003		2004-05	
Frequency of usage	narrowband	broadband	narrowband	broadband
More than once a day	41.7	74.7	33.5%	71.7%
Once a day	20.0	15.3	40.9%	16.6%
More than once a week	31.0	9.4	17.7%	10.1%
Once a week	6.3	0.3	7.9%	0.9%
Less	1.0	0.3	0.0%	0.8%
Chi ^{2}	317.67		116.13	
Sign. 2-tailed	0.00		0.00	
Session length	narrowband	broadband	narrowband	broadband
< 15 min.	4.8	0.5	1.8	1.2
15 min. – 1 hour	36.8	12.9	22.6	17.4
1 – 2 hours	34.0	30.0	41.5	32.7
2-4 hours	18.6	33.7	32.3	31.8
4-8 hours	4.5	14.7	1.8	11.7
> 8 hours	1.3	8.2	0.0	5.3
Chi ^{2}	298.71		36.74	
Sign. 2-tailed	0.00		0.00	
Internet experience	narrowband	broadband	narrowband	broadband
1 – 3 years	13.1	11.6	45.8	17.2
4 - 6 years	50.0	53.2	38.5	43.1
7 – 9 years	23.3	23.9	12.7	28.1
More than 9 years	13.5	10.9	3.0	11.6
Chi ^{2}	8.85		123.62	
sign. 2-tailed	0.840		0.00	
Internet experience	narrowband	broadband	narrowband	broadband
Laggard	0.2	0.2	0.6	0.1
Beginner	5.4	2.1	10.4	1.2
Average user	34.8	20.4	53.0	22.3
Experienced user	32.3	38.0	27.4	41.0
Very experienced user	15.3	25.5	5.5	23.6
Professional user	11.9	13.8	3.0	11.8
Chi ^{2}	81.92		141.22	
Sign. 2-tailed	0.00		0.00	

Table 12.4. Difference in annoyances between narrowband and broadband over the years (2003, 2004-05)

Annoyances (%)	2003				2004			
	nb	bb	chi2	sign. 2-tailed	nb	bb	chi2	sign. 2-tailed
Spam	60.8	**78.0**	67.07	0.00	49.4	**62.8**	16.69	0.00
Slow internet connection	**78.0**	40.5	249.01	0.00	31.1	20.5	5.08	0.08
High costs	**43.3**	17.2	165.95	0.00	**27.4**	12.6	27.07	0.00
Technical failure	18.4	**36.2**	64.84	0.00	12.8	20.5	3.49	0.18
Lack of privacy	7.7	**16.1**	26.12	0.00	5.5	6.9	2.79	0.24
Information overload	10.6	9.3	0.83	0.37	4.9	3.0	8.61	0.65
Surfing of family members	4.8	4.2	0.35	0.56	1.2	6.5	7.24	0.27
Pop-ups	-	-	-	-	38.4	**69.1**	33.34	0.00
Viruses	-	-	-	-	39.6	**58.4**	14.19	0.00
None	3.7	*6.0*	4.49	0.02	6.7	2.1	4.16	0.13

Italic = significant at 5% level. **Bold** *= significant at 1% level*

In the measurement of 2004-05, two categories were added: viruses and pop-ups. It turned out that many Internet users and especially broadband users suffered from both pop-ups and viruses. Although pop-ups and viruses are a hard problem to tackle it is, from different point of views, important to get this under control. It could also help the adoption of broadband Internet, as people could associate broadband with viruses, pop ups, and spam.

Thresholds and Benefits

Table 12.5 shows the thresholds for narrowband users to adopt a broadband connection. Until 2004, an important barrier to adoption was increased costs (both of installation and subscription). Respondents that did have broadband in the same years mentioned the reduction of costs as a reason to adopt broadband (Table 12.6). The paradoxical difference in these answers can be explained by the results from other parts of the research. Narrowband users are much lighter users of internet than broadband users (see Table 12.3). For light users there is no reason to invest in 'always-on' and 'flat fee.' For heavy users a broadband connection is relatively cheap. In the diary project similar results were found: the question is whether people are willing to pay rather than the absolute price. This willingness depends on the degree to which the internet has an important place in their lives.

After the measurement of 2003, we see a big change in this picture: the costs are no longer important thresholds to adopt broadband. This is logical when we consider the tremendous offers that Internet service providers did to Dutch consumers. Several ADSL and cable light versions were introduced at only a fraction of the costs that applied before.

The fear of the addictive character of broadband was in 2001 quite an issue and also reason to not adopt broadband. In later years, this was of virtually no importance.

A growing percentage of narrowband users are satisfied with the connection they now own. In 2003, 13.9 percent found their connection met their standards. In 2004-05, this percentage had risen to 41 percent. This implies that the people that still have a narrowband connection are satisfied with that, because they are light Internet users and see no immediate reason to change to broadband.

In 2003, an important reason that respondents do not have broadband appears to be a lack of availability of broadband. More than 90 percent of the respondents that state that a lack of availability is the reason they do not have internet also complain about the slowness of their internet connection, only 5 percent of this group find installation costs too high and almost 10 percent finds the subscription costs too high. This implies that these respondents in 2003 were willing to invest in a broadband connection. And they probably did in the following year: the availability of

Table 12.5. Thresholds for narrowband users to adopt broadband

Reasons for rejection	Narrowband 2003 (N= 555)	Narrowband 2004-05 (N=164)
Unsuitable PC	4.6	5.0
Not interested	5.4	-
Installation costs	28.2	8.2
Subscription costs	39.1	7.2
No added value	4.5	4.3
Satisfied with narrowband	13.9	41.3
Lack of skills	5.3	6.1
Addictive character of broadband	2.0	0.0
Unwanted information (sex. adds)	2.6	1.3
Not available	33.8	4.3
Too much hassle	3.6	3.7
Other	6.7	5.3

broadband had rapidly expanded in the Netherlands and in 2004 the availability was no longer an issue.

Only a small amount of narrowband users claim that lack of knowledge is the hurdle that keeps them from switching to broadband. One could argue that it does not take more skills to operate the internet via broadband than it does via narrowband. Maybe even the opposite is true: in contrast to narrowband users, broadband users perceive themselves as very experienced users, even if the have only had used internet for a short period of time.

The most important reasons to adopt broadband are high speed, controlling costs by the flat rate principle of broadband and the always-on aspect. These broadband features stay important reasons to adopt it over the years, as did the quality of sound of video and audio. The fact that with broadband one can keep telephone lines free for incoming and outgoing calls has also been an important reason for people to connect to broadband. The possibility to connect computers was also important in the decision to adopt broadband.

Table 12.6. Reasons for broadband users to adopt broadband

Reasons for adopting	Broadband 2003 (N= 1770)	Broadband 2004-05 (N=938)
Controlling costs by flat rate	53.9	41.8
Always-on	62.8	52.8
Tele working	4.8	6.6
Better sound and video quality	15.0	25.4
Social aspects: communication	3.5	3.5
Social considerations (mobility. environment)	0.8	0.2
High speed	63.5	69.6
School	4.9	4.9
Time saving	5.3	12.4
Connecting computers	10.9	14.2
Telephone lines free	33.6	27.4
Other	5.1	5.3
Online gaming	7.5	7.9
Sharing large files (film. music etc.)	9.7	15.8

Expected and Realised Cost Reductions

Since costs play such an important role in determining whether individuals switch from narrowband to broadband, we will elaborate on that further. In the questionnaires we asked narrowband users in what areas they expected to reduce costs when they would install broadband. Also, we asked users of broadband what cost reductions they have actually realised. Table 12.7 shows the results of this question.

Both groups state the biggest cost reduction lies in a decrease of communication costs. For people that have broadband this is even stronger than for narrowband users. This is constant over the years. In 2003, narrowband users expect a cost reduction for buying and renting of videos whereas broadband users state less realized cost reductions from that. In 2004-05, however, the realized costs by broadband users overweigh the expected costs of narrowband users. Surprisingly, narrowband users do not expect as many cost reductions as they did in 2003 (45 percent says to expect no cost reductions. whereas in 2003 this was approximately 27 percent). This could be due to the fact that the people that did expect cost reductions now have made the step to broadband. The light users that still have narrowband connections will indeed not realize so many cost reductions, simply because they do not have that big an interest in downloading films or music.

Differences in Patterns of Use

Table 12.8 shows an overview of the usage patterns of narrowband users in comparison to broadband users and the differences between figures from 2003 and

Table 12.7. Expected and realised cost reductions (2003, 2004-05)

Cost reduction (%)	2003		2004-05	
	Expected (N=555 narrowband)	**Realised** (N=1770 broadband)	**Expected** (N=164 narrowband)	**Realised** (N=933 broadband)
Communication cost	47.6	58.6	39.1	41.2
Music	21.0	31.0	7.2	30.6
Video	16.7	10.4	3.1	8.6
Software	20.4	26.2	4.7	20.3
Travel cost	-	12.9	15.6	10.6
None	27.0	19.5	45.5	25.0
Do not know	26.6	12.0	-	14.0

2004-05. The most striking result when looking at this table is that the differences are relatively small. Of all the services that were presented only 10 services differed significantly at the 1 percent level (using chi^2, which is relatively sensitive to a large sample size). In 2004-05 more significant results (18) were found.

When we look at information activities we see that in 2004-05, broadband users more often used online reference works, such as telephone books and maps, than narrowband users (48.7 percent compared to 22.6 percent for narrowband users). Also, information via sound and video is much more important to broadband users than to narrowband users and it is increasingly important to broadband users. Offering information through their own websites is also getting more popular for broadband users.

In communication, we see that chatting is becoming more popular especially for broadband users but also for narrowband users. Reading weblogs is significantly done more often by broadband users, although not frequently done (8 percent). Discussion forms are more of a thing for broadband users, in 2003 as well as in 2004-05. Webcams for communication were not included in the 2003 measurement, but are significantly more used by broadband users than by narrowband users (12 percent compared to 5.9 percent).

In entertainment activities, the image is steady over the years: broadband users more often

play online games, downloads or watch films, TV, video recordings, download, or listen more often to music via the internet. Fun mail and fun surfing are important to narrowband users.

The most changes we see are in transaction activities. In 2003, there were no significant differences between narrowband and broadband users, but in 2004-05, there were four: broadband users more often buy products and services online, buy or sell through online auctions, do telebanking, and make reservations than narrowband users.

Discussion and Conclusion

The most important results of the two surveys can be summarised as follows:

- There are no dramatic differences in user characteristics between broadband and narrowband users. The biggest difference lies in gender differences, but these characteristics remain steady over the years.

Table 12.8. Differences in usage patterns of narrowband and broadband (2003, 2004-05)

% respondents that uses an application								
	2003				**2004**			
Information	**nb**	**bb**	**chi²**	**sign.**	**nb**	**bb**	**chi²**	**sign.**
Info through search engine	77.7	80.4	2.58	0.06	87.8	89.1	0.23	0.64
Info through portals	*38.2*	32.1	7.07	0.01	31.1	35.4	1.12	0.29
Info through reference work	19.8	20.5	0.18	0.73	22.6	**48.7**	38.42	0.00
Info through sound and video	7.2	**13.9**	19.10	0.00	7.3	**40.8**	68.16	0.00
Info through application form	**13.8**	8.3	15.24	0.00	8.5	10.9	0.84	0.36
Info via subscription	10.2	11.0	0.35	0.50	2.4	**11.8**	13.1	0.00
Info through discussion group	6.5	*9.5*	5.19	0.02	10.4	11.6	0.20	0.65
Info through own website	9.0	11.3	2.51	0.13	3.0	**16.3**	19.94	0.00
Communication	**nb**	**bb**	**chi²**	**sig.**	**nb**	**bb**	**chi²**	**sig.**
Chat (ICQ)	32.8	**52.4**	70.13	0.00	14.9	**85.1**	80.39	0.00
Chat (anonymous/website)	9.2	11.4	2.37	0.07	0.6	*3.9*	4.51	0.03
Internet telephony	2.5	2.2	0.16	0.75	3.0	4.8	1.00	0.32
Videophone	0.8	1.9	3.38	0.07	0.6	1.0	6.19	0.66
Web log (read)	4.7	3.9	0.71	0.41	3.7	**8.0**	15.15	0.00
Web log (write)	2.7	1.8	1.83	0.18	1.2	2.7	1.23	0.26
Discussion forum	9.5	**16.9**	19.56	0.00	10.4	**15.6**	3.08	0.08
SMS/MMS	11.1	11.2	0.00	1.00	2.4	4.7	1.72	0.19
E-mail	96.5	95.0	1.33	0.29	90.2	94.0	3.21	0.07
Webcams	-	-	-	-	5.9	**12.0**	8.63	0.00
Entertainment	**nb**	**bb**	**chi²**	**sig.**	**nb**	**bb**	**chi²**	**sig.**
Online games	18.0	**30.8**	43.84	0.00	6.7	**28.0**	33.83	0.00
Download/watch movie	7.4	**26.0**	97.50	0.00	1.2	**17.0**	27.88	0.00
Upload movie	0.9	1.2	0.46	0.65	0.6	1.0	0.19	0.66
Set up or maintain community	6.1	6.4	0.34	0.63	1.2	*4.7*	4.25	0.04

continued on following page

Table 12.8. continued

	nb	bb	chi²	sig.	nb	bb	chi²	sig.
Share in communities	7.2	8.6	2.02	0.16	12.8	*6.6*	7.56	0.01
Watch television	2.9	**10.1**	32.14	0.00	4.3	**13.0**	10.25	0.00
Watch video recordings	15.1	**23.3**	22.43	0.00	7.9	**18.6**	11.34	0.00
Upload video recordings	1.8	0.7	4.25	0.05	0.0	0.2	0.35	0.55
Download or listen to music	40.8	**66.5**	145.57	0.00	9.1	**47.3**	83.45	0.00
Upload music	2.7	*4.9*	62.17	0.01	1.2	2.0	0.49	0.48
Download or looking at pics	*33.3*	25.6	6.70	0.01	18.3	*25.6*	4.04	0.04
Upload pictures	**11.3**	5.5	17.95	0.00	6.7	7.0	0.01	0.90
Fun mail	**41.0**	31.5	9.31	0.00	**50.0**	23.2	50.82	0.00
Fun surfing	70.7	64.2	0.64	0.44	54.3	54.6	0.01	0.94
Transaction	**nb**	**bb**	**chi²**	**sig.**	**nb**	**bb**	**chi²**	**sig.**
Buy products/services (retail)	57.1	54.4	0.19	0.67	35.4	**55.7**	23.21	0.00
Buy or sell marketplace	17.8	18.0	0.13	0.75	15.9	*23.8*	5.62	0.02
Buy or sell auction	10.9	*14.7*	5.98	0.01	4.9	**14.7**	11.69	0.00
Swap products	3.3	3.5	0.11	0.89	1.2	*5.1*	4.94	0.02
Telebanking	73.3	*77.0*	6.33	0.01	61.0	**81.0**	32.72	0.00
Reservations	34.7	32.7	0.19	0.68	18.9	**30.5**	9.22	0.00

Italic = significant at 5% level. **Bold** *= significant at 1% level*

- The image of broadband users in 2003 is quite similar to that of broadband users in the following year. Broadband users are significant heavier users of the Internet.
- Spam, pop-ups, and viruses are important annoyances for especially broadband users.
- Until 2004, important barriers to adoption were increased costs (both of installation and subscription). In the later measurement we see a big change in this picture: the costs are no longer important thresholds to adopt broadband, because of offers that ISPs make. Also, high costs are becoming a less important annoyance for both groups of users.
- In 2003, an important reason that respondents did not have broadband is a lack of availability of broadband, the availability of broadband had rapidly expanded in the Netherlands so that in 2004 the availability was no longer an issue.

- Broadband characteristics, such as flat fee (controlling costs), always-on and high speed are the most important reasons to adopt a broadband connection.

- Differences in patterns of use between narrowband and broadband users are getting more noticeable. In entertainment activities the differences already were evident in the 2003 measurement, but more differences are now also found in information, communication and transaction activities.

This research shows that it is important to measure the adoption of new technologies such as broadband. Already in one year we see changes in the reasons and thresholds for people to adopt and in differences between broadband and narrowband users. This has to do with developments in the market (better availability, cheaper offers, and more services). Currently we see a steady growth in the number of broadband subscriptions. Ubiquitous availability of broadband will further increase the number of subscriptions. However, this research also shows that not everybody is keen on taking the step from narrowband to broadband. Only when broadband is standard (like colour TVs nowadays) and prices do not differentiate, also the lighter users will take the step to broadband. Especially for this group, that does not appreciate the benefits of the Internet strongly, costs are an important factor.

The use of technologies, such as broadband, actually is the subtraction of the technological possibilities and the needs people have. When we look at our theoretical framework we can draw the following conclusions. A broad range of theories that describe the adoption and use of new technologies is available. Diffusion of innovations shows us how a relatively small group of innovate users stimulate further development of internet and broadband technology, while other groups follow developments from a distance. Uses and gratifications show how there is a clear relation between people's motivations and the actual use of technology. With regard to media choice theory and the technology acceptance model it is possible to state that people seek themselves a route with little resistance: they simultaneously try to maximize results and minimize costs.

All this illustrates how the adoption and use of new technologies is a complex process. In our opinion there is not so much a need for new theories, but rather a need for a framework that grasps the complexity of the adoption and use process. This chapter has provided an initiative towards integrating several theories that look at technology from a user perspective. Users can tell us what they do with new technologies, why they do it and how technology changes our lives.

As with all research there are some limitations with regards to this research, however, one of the limitations of this research offers opportunities. While these results are solely based on Dutch Internet users, the chance of comparing results from similar survey conducted in other countries, would give further insights into the adoption of technologies such as broadband. A second limitation involves the sample. Because of the convenience sample used, the segmentation of demographical characteristics may not be fully representative. In future in-depth research on the relationship between demographical characteristics and technology adoption this should be addressed.

References

Anderson, B., Gale, C., Jones, M.L.R., & McWilliam, A. (2002). Domesticating broadband-what consumers really do with flat rate, always-on and fast Internet access. *BT Technology Journal, 20*(1), 103-114.

Andriessen, J.H.T.H. (1994). Conditions for successful adoption and implementation of telematics in user organizations. In J.H. Erik Andriessen & Robert A. Roe (Eds.), *Telematics and work*. Hove, Sussex: Lawrence Erlbaum.

Bandura, A. (1986). *Social foundations of thought and action: A social cognitive theory*. Englecliffs, NJ: Prentice Hall

Bunz, U. K. (2001). *Usability and gratifications: Towards a website analysis model*. Presented at the National Communication Association Convention. Atlanta, GA.

Chua. S.L., Chen. D.T., & Wong, A.F.L. (1999). Computer anxiety and its correlates: a meta-analysis. *Computers in Human Behavior, 15*, 609-623.

Davis, F.D. (1989). Perceived usefulness. Perceived ease of use, and user acceptance of information technology. *MIS Quarterly, 13*(3), 319-341.

Daft, R.L., Lengel, R.H. (1986). Organizational information requirements. Media richness and structural design. *Management Science, 32*(5), 554-571.

Daft, R.L., Lengel, R.H., & Trevino, L. K. (1987). Message equivocality. media selection and manager performance: Implications for information systems. *MIS Quarterly, 11*(3), 354-366.

de Haan, J., & Steyaert, J. (2003). *Jaarboek ICT en samenleving 2003. De sociale dimensie van technologie*. Amsterdam: Boom.

Durndella, A., & Haag, Z. (2002). Computer self efficacy. computer anxiety. attitudes towards the Internet and reported experience with the Internet, by gender, in an East European sample. *Computers in Human Behavior, 18*(2002), 521-535.

Firth, L., & Kelly, T. (2001). *The economic and regulatory implications of broadband*. Geneva, ITU. Retrieved August 5, 2005, from http://www.itu.int/osg/spu/ni/broadband/workshop/briefingpaperfinal.doc

Fulk, J., Schmitz, J., & Steinfield, C. (1990). A social influence model of technology use. In J. Fulk & C. Steinfield (Eds.), *Organisations and communication technology* (pp. 117-140). Newbury Park: Sage.

Galbraith, J.J. (1973). *Designing complex organizations*. Reading MA: Addison-Wesley.

Hill, T., Smith, N.D., & Mann, M.F. (1987). Role of efficacy expectations in predicting the decision to use advanced technologies: The case of computers. *Journal of Applied Psychology, 72*, 307-313.

Igbaria, M., & Iivar, J. (1995). The effects of self-efficacy on computer usage pergamon. *International Journal of Managemet Science, 23*(6), 587-605.

Katz, E., Blumler, J. G., & Gurevitch, M. (1974). Utilization of mass communication by the individual. In J. G. Blumler & E. Katz. (Eds.), *The uses of mass communications. Current perspectives on gratifications research*. Beverly Hills, CA: Sage.

Katz, J. E. & Rice, R. E. (2002). *Social consequences of Internet use: Access, involvement, and interaction*. Cambridge, MA: MIT Press.

Katz, E., Gurevitch, M., & Haas, H. (1973). On the use of the mass media for important things. *American Sociological Review, 38*, 164-181.

Katz, J. E., Rice, R. E., & Aspden (2001). The Internet, 1995-2000 access, civic involvement, and social interaction. *American Behavioral Scientist, 45*(3), 437-456.

Lin, C.A. (2002). Perceived gratifications of online media service use among potential users. *Telematics and Informatics. Volume, 19*(1), 13-19.

Organization for Economic Cooperation and Development. *The development of broadband access in OECD countries*. October 29, 2005.

Papacharissi, Z., & Rubin, A.M. (2000). Predictors of internet use. *Journal of Broadcasting and Electronic Media, 44*(2), 175-197.

Pociask, S. (2002). *Building a nationwide broadband network: Speeding job growth*. White paper prepared for New Millennium Research Council by TeleNomic Research. Retrieved from http://www.newmillenniumresearch.org

Rice, R.E. (1993). Media appropriateness: Using social presence theory to compare traditional and new organizational media. *Human Communication Research, 19*(4), 451-484.

Rogers, E.M. (1983). *Diffusion of innovations*. New York: Free Press

Rosengren, K. E., Wenner, L. A., & Palmgreen, P. (1985). *Media gratifications research: current perspectives*. Beverly Hills: Sage.

Savolainen, R. (1999). The role of the Internet in information seeking. Putting the networked services in context. *Information Processing & Management, 35*(6), 765-782.

Short, J., Williams, E., & Christie, B. (1976). *The social psychology of telecommunications*. London: Wiley.

Silverstone, R., & Haddon, L. (1996). Design and the domestication of information and communication technologies: Technical change and everyday life. In R. Mansell & R. Silverstone (Eds.), *Communication by design. The politics of Information and Communication Technologies.* Oxford: Oxford University Press.

Sitkin, S.B., Sutcliffe, K.M., & Barrios-Cholin, J.R. (n.d.). A dual capacity model of communication media choice in organizations. *Human Communication Research, 18*(4), 563-598.

Steinfield, C. (1986). Computer-mediated communication in an organizational setting: explaining task-related and socio-emotional uses. In M. McLaughlin (Ed.), *Communication yearbook 9.* Beverly Hills, CA: Sage.

Tan, A. (1985). *Mass communication theories and research.* New York: John Wiley & Sons.

Thierer, A. (2002). Solving the broadband paradox. *Issues in Science and Technology, 18*(3), 57-62.

Trevino, L. K., Daft, R. L., & Lengel, R. H. (1990). Understanding managers' media choices: A symbolic interactionist perspective. In J. Fulk & C. Steinfield (Eds.), *Organisations and communication technology* (pp. 71-94). Newbury Park, CA: Sage.

Trevino, L.K., Webster, J., & Stein, E.W. (2000). Making connections: Complementary influences on communication media choices, attitudes and use. *Organization Science, 11*(2), 163-182.

Venkatesh, V., & Davis, F.D. (2000). A theoretical extension of the technology acceptance model: Four longitudinal field studies. *Management Science, 46,* 186-204.

Wales, C., Sacks, G., & Firth, L. (2003, January 18-23). *Killer applications versus killer attributes in broadband demand.* PTC Conference, Hawaii.

Webster, J., & Trevino, L. K. (1995). Rational and social theories as complementary explanations of communication media choices: Two policy capturing studies. *Academy of Management Journal, 38,* 1544-1572.

Wellman, B., Quan-Haase, A., Witte, J., & Hampton, K. N. (2001). Does the Internet increase, decrease, or supplement social capital? Social networks, participaiton, and community commitment. *American Behavioral Scientist, 45*(3), 437-456.

Chapter 13

Broadband in Dutch Education:
Current Use, Experiences, and Thresholds

Karianne Vermaas, Dialogic/Utrecht University, The Netherlands

Sven Maltha, Dialogic Innovation & Interaction, The Netherlands

Abstract

Broadband potentially has benefits for education, but in order to be beneficial it has to be used. In this chapter, we have investigated from a user perspective: (1) to what extent broadband is used in Dutch education (in the classroom as well as in the organisation as a whole); (2) the experiences teachers have with broadband, including impediments and added value. This was done by a survey under 221 Dutch teachers, ICT-coordinators, and school boards. Results show that teachers, ICT coordinators, and school boards are interested in using broadband in their schools as they see the added value, but there seems to be an impasse: without infrastructure, there are no services and without services there is no need for infrastructure. Schools can break out of the causality dilemma by giving an impulse to the market by combining forces and demand. Moreover, teachers need to be trained in using the new tools and service.

Introduction

The benefits of information technology for education are widely recognized. Different researchers describe a variety of benefits of information technologies for learning, such as increased access to learning opportunities, access to more and better information resources, availability of alternative media to accommodate different learning strategies, increased motivation to learn, and possibilities for both individualized learning and collaboration in learning (Niemi & Gooler, 1987). Furthermore, information technologies can provide access to enormous quantities of information available through Internet and online databases, reduce the limits of time and space for educational activities, enable self-paced learning, makes the teaching and learning enterprise more outcome-oriented, which enhances the ability of institutions to stimulate experimentation and innovation, and increases learning productivity (Massy & Zemsky, 1995).

All these benefits are also applicable to broadband. This infrastructure allows new forms of learning and can expose students to innovative forms of learning and new content that is not accessible with a narrowband connection. New forms of learning that need a fair amount of bandwidth often involve multimedia educational content. This allows the student to actually see, hear, and use the content to be learned (Roden, 1991). Students are able to learn by making short video clips, games, or ask questions to different experts via a videoconference system. Exposure to new forms of learning and new content can have a positive effect on the motivation of students to learn. Research even shows that video stimulates gaining of knowledge and aids retention and recall (Duchastel & Waller 1979; Goodyear & Steeples 1998; Mayer & Gallini 1990; Pahad, 1998; Shepard & Cooper 1982). Dutch students for example had a videoconferencing session with students from Palestine. Teachers were thrilled because students picked up the learning content in a very natural way and retained it for much longer. It made more impact on the students than learning it from a book (Vermaas, 2005).

Another benefit of broadband is that it can facilitate and enhance collaboration within and across institutions. For example, broadband can be used to share resources. Students can work together when they are inside the school, but also at home. Even students and teachers from different schools can work together, via videoconferencing or by sending (video) material that requires broadband. Moreover, students and teachers can collaborate with other institutions, such as libraries.

A specific advantage of broadband is that it possibly delivers efficiencies and cost reductions in administration and reporting (Broadband Stakeholder Group, 2003; Underwood et al.). Also, automating the administration and management of educational institutions can lead to cost reductions. The management of computers and the infrastructure itself can also be centralised, even outside of the school, which will be a relief for many schools that often have teachers or volunteers managing the infrastructure and information and communications technology (ICT).

Whilst the potential benefits of broadband are recognized, there are some impediments to its successful adoption. These range from a lack of experience or skills to high costs for roll out of broadband and subscription (Dialogic 2005a; NOIE, 2002; Wijngaert, Vermaas, & Maltha, 2003). This research provides more insights in these impediments from a user perspective.

Among the parties involved in Dutch education and (broadband) Internet there is an increasing need for research unravelling the status of the usage of, experience with and attitudes towards broadband in schools. Therefore this research into broadband for educational purposes was conducted. This research will be repeated, so the developments can be monitored over the years.

Current Subscriptions to Broadband and Applications

Before presenting any results, it is important to get an understanding of what is meant by broadband Internet in this research. Since the Internet was made available to the public in the early nineties through the World Wide Web, people have been able to go online with modems in combination with an analogue telephone line. Such a connection is now referred to as a narrowband connection. Broadband has not just one definition and the definition of broadband is changing as speeds change rapidly.

In this research the focus is not on the technology itself, therefore certain properties are given to distinguish broadband from the 'traditional dial-up connection.' In this research, the term narrowband is used for any connection that is established through dial-up access (traditional modems and an analogue telephone line). Therefore public switched telephone network (PSTN) and also integrated services digital network (ISDN) connections are considered narrowband. Broadband in general is used to indicate telecommunication in which information can be transmitted over a wide band of frequencies. Because of the wide band of frequencies that is available, information can be sent on many different channels within the band simultaneously, allowing more information to be transmitted in a given amount of time.

Two important characteristics of broadband Internet are 'flat fee' and 'always-on.' 'Flat fee' means that the subscribers pay a fixed amount of money per month, regardless of the actual time spent online (as opposed to dial-up access, where people pay per time-unit that they are connected). 'Always-on' means that there is direct connection at any time; there is no need to dial up. Based on these characteristics asymmetric digital subscriber line (ADSL), cable, and fibre optics are considered broadband in this research.

In the Netherlands the penetration of broadband in households is the highest in Europe (OECD, 2005), but the number of schools connected to a high speed infrastructure are not as high. Therefore the Dutch Ministry of Education, Culture, and Science has provided a financial aid for schools to roll out and connect to a fibre optics network. Within the forthcoming months approximately six hundred schools will be connected to a fibre optic network.

The next figure shows the status of Internet connections in Dutch education (ISP wijzer, 2005). It shows that ADSL (68 percent) is rising. In 2004, less than 58 percent of the schools had an ADSL connection. The number of PSTN and ISDN connections dropped considerably. Almost half of the schools in primary education and secondary educations use a connection with a download-capacity of 1-4 Mbps. The mean is about 3.3 Mbps. In 2004, this was less, namely 3.0 Mbps. In conclusion, more and more schools are using broadband and the mean bandwidth is increasing.

Many broadband services for educational purposes are being developed in The Netherlands. In the following frame, some Dutch good practices are described. The described good practices show that broadband can be employed in education in many different ways. It is roughly divided into information and communication. Part of the services is concerned with providing (existing or new) images. Other services are focussed on developing of video material by students or the ability of putting together educational material by students. Another important aspect of many educational broadband services is communication; for example communication between students and students and experts. Sometimes this communication goes beyond the Dutch border; students can present their material for example on international conferences or speak or collaborate with students from other countries.

Table 13.1. Infrastructure in schools (Source: ISP wijzer, based on TNS Nipo, 2005 (2004: n = 485, 2005 = 578)

	Total		Primary Education		Secondary Education	
	1 - 1 - 2004	1 - 1 - 2005	1-1-2004	1-1-2005	1-1-2004	1-1-2005
ADSL	58	68	64	75	34	32
Cable	24	19	23	17	28	29
Other DSL	4	4	1	1	19	20
ISDN	7	3	8	3	2	1
PSTN	1	1	1	1	1	0
Wireless	1	0	0	0	2	1
Other	5	7	1	4	19	21

For the services, videoconferencing (for communication) and streaming video (information) are much used applications. Many of the discussed service and applications are still in pilot phases and not yet available to every school in The Netherlands. Schools that do not have a broadband infrastructure cannot yet use and experience these services and applications.

Videoconferencing with an Expert (expert op afstand)

A number of schools are already working with videoconferencing with experts. For example, the possibilities of extra terrestrial life were discussed with a scientific journalist and with Palestinian students. Stories were exchanged and a scientist was asked several questions about migration in the United States. In other sessions, students could learn more about certain traits by asking professionals about it, such as car mechanics and nurses.

The Legend of the Seven Seas

In the legends of the seas secondary education students, in teams, made Web-based games. They are supported by TeleTop (electronic learning environment developed by Twente University). This activity replaced parts of lessons in information science, mathematics, and history. Even teachers were surprised about the enthusiasm and inventiveness of the students. Because of the enthusiasm, they go beyond the official teaching material. Students learn to work together and understand what is involved in developing such a product. Results were presented at the iEARN world conference in Senegal in 2005. Broadband is a condition for fast and problem free connection in order to show and play the game in a good way.

Schooltv Image Database (Schooltv beeldbank)

The Schooltv database is free and easily accessible. It contains many educational video clips. Every clip can be used directly in the class room. But the student can also use it at home to look things up to use in a paper or assignment. The subjects are suitable for primary education up to secondary education. The database is filled in cooperation with teachers. Broadband is necessary to display the high quality video and audio clips.

Discover net (Ontdeknet)

Discover net is an electronic learning environment. Via Discover net, students can get into contact with all sorts of experts from society. Also companies, organisations, government, and citizens can share information with the educational environment in a simple way. Everybody can serve as an expert, as long as that person has expertise on every imaginable terrain. The leading figure is an agent Onty, the Ontdekvis (Discover fish) that leads students through the application and information. Students work together on projects (virtual papers and projects) about different subjects. Students receive knowledge and material (video, sound, text, images), via the Web, of linked experts. Although Discover net is available in a narrowband version, the properties of the concept are more noticeable with broadband (especially the quality of media files).

Expose Your Talent

In this video competition for secondary education, groups of students make short films. In workshops, the students are supported with scenario writing, camera operation, and editing. Teachers are assisted and learn how to use and integrate making movies and streaming media in the class room. The second edition of the competition was held in 2006. Students and teachers are very enthusiastic about this application (Vermaas, 2005).

Kennisnet-Videoportal

This video portal is comparable to an online video library for education. In the video library, teachers and students can look for and flip through a large amount of video material. The video material can also be saved and displayed in the portal.

Teleblik

With Teleblik (operational since late 2005) a myriad of audio-visual material, such as old TV news, educational TV, and archive material is made accessible in a safe, categorised way and free of copy rights. It is meant for primary education, secondary education, and vocational education and training and adult education (VET). A search engine makes the material accessible in a fast and easy way. There are also suggestions on how to use the material in the classroom. This application requires broadband, because video material can only be played in a good quality with the necessary band width.

Learning Circles with Power Users

In learning circles students communicate, discuss, and publish in well defined periods with each other online about global issues, such as the Millennium Goals. Existing learning circles are extended to 'learning networks,' which will exist for at least two years. With research, the added value will be evaluated of this new form of learning: collaborative learning and informal learning and the use of broadband applications to support this. The used applications will only work smoothly with a fair amount of band width.

Theoretical Framework

Added Value of Broadband for Education

Broadband for education is still in its infancy and elaborate conceptual models to explain the (non) adoption and usage are not available. We can, however, find important clues in literature on adoption of new technologies. Perceived benefits and relative advantage are important in the adoption of technologies (Davis, 1989; Rogers, 1995). This is of course also the case with the adoption of broadband for educational purposes. Relative advantage (Rogers, 1995) includes "economic profitability," low initial cost, increase of comfort, social prestige, saving of time and effort, and immediacy of reward." Translated to the relative advantage of broadband in education, we can sum up the following benefits for a school as a whole:

Table 13.2. Added value for the school as a whole

Cost reductions (telephony, Internet subscription bills)	low costs
Internal process support (e.g., administration)	increase of comfort
Shared use and management of ICT services and applications	increase of comfort
Internal educational applications (better and variation in teaching material)	increase of comfort, social prestige, increase of comfort
External communication with students (e.g., via email)	social prestige, increase of comfort
External communication with parents (e.g., via website)	social prestige, increase of comfort

Table 13.3. Added value for the in-class and class related activities

Facilitating individual learning
Facilitating and stimulating location independent learning
Supporting the development of individual skills
Facilitating class activities
Stimulating and facilitating collaboration between students
Stimulating and facilitating interaction between teacher and student

More specifically, broadband can have advantages and benefits for in-class activities of lesson related activities. Several Dutch action programmes (e.g., "learning with ICT" of the Dutch Ministry of Education, Culture, and Science) concerned with ICT and education stress the importance of individual learning, location and time independent learning, development of individual skills of students, class activities and teamwork and interaction between teachers and students. These aspects are also mentioned by researchers such as Niemi and Gooler (1987) and Massy and Zemsky (1995).

On the one hand the possible added value of broadband is recognized by many researchers. There are however, some thresholds to overcome before adoption and widespread use of broadband in education is a fact. These thresholds are presented in the following section.

Thresholds

For a technology to be adopted there has to be a fit between the task and the technology or medium. This has often been the central issue in many media choice theories. The social presence concept (Short, Williams, & Christie, 1976), media richness (Daft & Lengel, 1986; Trevino, Daft, & Lengel, 1990), the social influence Model (Fulk, Schmitz & Steinfield, 1990), the dual capacity model (Sitkin, Sutcliffe, & Barrios-Choplin, 1992), and media appropriateness (Rice, 1993) offer comparable starting points for the analysis of media choice. The basic assumption is that a good task/medium fit is essential for effective communication. A lack of the fit between task and medium or technology can cause rejection. Apart form the task technology fit (TTF) there are other conditions for adoption and use of a medium or technology. There are several barriers (Bouwman et al., 1996) or thresholds (Van Dijk, 1997) to be overcome before an individual will adopt a medium or technology. Van de Wijngaert (2001, 2004) in her 3-thresholds model discerns physical accessibility, affective accessibility and fit of the medium or technology. Bouwman et al. (1996) distinguish four thresholds. Apart from physical accessibility, they treat financial, technical and cognitive accessibility.

Accessibility

After determining the TTF, the question arises: which media are accessible to someone in a certain situation? Accessibility can be defined as the subjective perception of the user of the degree in which he will encounter difficulties while using the technology (Auster & Choo, 1993). In order to use a medium or technology a user has to have access to it, by owning it or by accessing it in a different way (for teachers to use broadband in the classroom there has to be a broadband infrastructure in the school). This actual access is referred to as physical accessibility. This could be related to a financial accessibility. This is the direct price of the investments to use a medium or technology, such as costs for devices and subscriptions. For many schools in The Netherlands it was too expensive to switch to broadband. But, with several ADSL offers (ADSL was even offered to schools for free) and lately a fair amount of price reductions for fibre optics (gained by demand bundling), this is changing rapidly.

Technical accessibility is concerned with the way the technology is offered and the consequences it has for the possibilities to use. Important aspects are user friendliness and the user interface. Cognitive accessibility is related to the technical accessibility. It involves the amount of experience that is needed to use a medium. This is related

Table 13.4. Thresholds for adopting broadband in the school (school boards and ICT managers)

Willingness	No need	Current infrastructure is sufficient It does not have added value in the class
Ability	Physical barrier	No broadband in the area of the school
	Financial Barrier	High subscription costs High costs for additional and new material and devices
	Cognitive barrier	Lack of technical skills and experience Impossible or hard to integrate in educational programme
	Technical barriers	Unsuitable devices, such as old PCs, etc. No or no accessible broadband content such as streaming media Not enough devices such as PCs, beamers, etc. Technical problems

Table 13.5. Thresholds for adopting broadband in the classroom (teachers)

Willingness	No need	The current infrastructure is sufficient, there is no need There is no added value for the classes given
Ability	Physical barrier	Lack of educational broadband content impossible or hard to integrate in educational programme (policy developed by school board)
	Financial Barrier	Not to be expected since the teacher does not have to pay for using broadband applications in the class room
	Cognitive barrier	Lack of skills
	Technical barriers	Lack of necessary equipment (such as PCs) Unsuitable equipment (old PCs, etc.)

to what Davis (1989) refers to as perceived ease of use: the degree to which a person believes that using a particular system would be free from effort.

The possible thresholds for the adoption of broadband in Dutch education are depicted in Table 13.2. This table also clarifies that there are thresholds of *willingness* and thresholds of *ability*; if there is no need, if the technology does not fit the task, they will be *unwilling* to use the technology. If, on the other hand, there is no broadband available, one has not enough money to buy or use the technology, has not enough skills to use it, or has no personal computer (PC) or other devices to use the technology, there is no *ability* to adopt.

Research Goal, Questions, and Methods

The goal of this research is to give an overview of whether and how broadband is used in Dutch education and what the experiences (both positive and negative) of users are. This is all investigated from a user perspective as the users of the technology, such as teachers, ICT coordinators, school leaders, and students play an important role in the success or failure of broadband in education. The research questions are:

- To what extent is broadband used in Dutch education (in the class room as well as in the organisation as a whole)?

- What are the experiences teachers have with broadband, including thresholds and added value?

Table 13.6. Classification of educational broadband applications available

| Short video clips/ animation via Internet |
| Online photos/drawings/maps |
| Online audio material |
| Complete productions, such as documentaries |
| Educational online games |
| Online collaboration tools, such as electronic learning environments |
| Creating and provide educational broadband content |
| TV via Internet |
| Videoconferencing |
| Chat / MSN |

In order to measure the development of broadband use and experiences in primary education, secondary education, and vocational education and training and adult Education (VET) an online questionnaire research was conducted under Dutch teachers, ICT-coordinators, and school boards.

Because this research area is rather new, we have invested effort in pre-testing our survey within different schools and it was evaluated by three parties involved in broadband in Dutch education. Also, because of the newness of this matter, there is no real classification of broadband applications for education. Together with the aforementioned parties we have developed the following classification.

Data Collection and Response

Data collection was between May until July, 2005, resulting in a total of 221 respondents. Most respondents were from secondary education (59 percent). Twenty-nine percent of the respondents work in primary education and 11 percent in VET. Most of the respondents are teachers (49 percent) and ICT-coordinators (46 percent). The remaining 5 percent are school boards.

Respondents received an e-mail with a link to the questionnaire. Also, there were links on various official Dutch educational websites, in newsletters, and in mailing lists. While reading and interpreting the results of this research, one has to bear in mind that the group respondents are rather early adopters. The results are therefore not directly representative for all teachers, ICT coordinators and school boards, but the results do give a good impression of what this group does with this infrastructure and which problems and difficulties they encounter.

Results

Equipment and Infrastructure

Before focusing on the usage of and experiences with broadband, we give a brief overview of the equipment and infrastructures in the schools. More than 62 percent of the schools have 20 or more personal computers that can be used by students during classes. This is however, not the same for every school type. Especially in VET there are often more than 20 PCs (88 percent) for class purposes. These are often large schools with many locations. In primary education there are often less computers; 29 percent has one to five PCs that can be used during classes.

In 97 percent of the cases the PCs are also connected to the Internet. Here, no big differences between education types are found. Apart from PCs, almost all schools have a network within one location. These results are comparable with results from the national ICT-monitor 2004-2005 (ITS/IVA, 2005). Mostly it involves fixed networks (75 percent), whereas 17 percent a wireless network uses. If a school has more than one location, these are in more than half of the cases linked by a network (42 percent by a fixed network and 18 percent by another network). 30 percent of the schools with more locations do not have a network that links the locations.

To get an impression of what teachers categorize under broadband, a perception question was added in the survey. Most respondents think of broadband in terms of fibre optics (35 percent) or ADSL (32 percent). For 27 percent, cable access also meets broadband criteria. Most schools in this research have an ADSL (39 percent) or cable (26 percent) infrastructure. Fibre optics occupies a third place (17 percent). The last group is larger than we see in the rest of The Netherlands (ISP wijzer, 2005).

Impediments for Switching to New Broadband Infrastructure

Many schools in The Netherlands use an ADSL infrastructure, because one provider offers it to the schools for free. This offer will expire in 2007 and schools have to make a decision on their future infrastructure. More than half of the schools are not planning on switching to a new broadband infrastructure in the near future (51 percent). Ten percent of the schools has ideas or wishes to do so and 6 percent have actual plans to make a switch. In primary education teachers, school boards and ICT coordinators are most reluctant to make a switch; 74 percent say 'no.' In secondary education, there is al lot of uncertainty; 47 percent of the ICT coordinators say they do not know whether there are plans, wishes, or ideas.

Table 13.7. Reasons for not switching to a new broadband infrastructure (more than one answer allowed)

Current infrastructure is sufficient	23%
High subscription costs	12%
High costs for additional and new material and devices	8%
Lack of technical skills and experience	4%
It does not have an added value in the class	4%
Unsuitable devices, such as old PCs etc.	3%
No or no accessible broadband content such as streaming media	2%
Not enough devices, such as PCs, beamers, etc.	2%
Technical problems	1%
Do not know	34%
Other	7%

The question is what the most important reasons are for schools to not want to switch to a faster broadband infrastructure (Table 13.7). What is remarkable is that most respondents do not know what the reason is that the switch to a new broadband connection is not made. Apparently they are not involved in the decision making process. The three most important reasons for not switching to a new broadband connection are: the current infrastructure is sufficient (23 percent), high subscription costs (12 percent), high cost for required devices and material (8 percent). Respondents do not worry about technical problems or lack of devices such as PCs and beamers.

The fact that many schools in primary education are not planning on switching to other infrastructure is probably related to the fact that these schools are financially under-privileged compared schools in VET or secondary education.

Usage of Broadband Services in the Classroom

Internet is extensively used in the class room; 41 percent of the teachers use the Internet more than once a day in the class room and 19 percent a few times a week. There are no big differences between school types with regard to the frequency of usage. The use of Internet in the class room in many cases results in less usage of traditional media, such as slides, VCR, and Television. Traditional media are less often used (25 percent) or a lot less often (18 percent) by teachers since they use Internet in the class room. 40 percent of the teachers uses other media just as often as they did before starting to use the Internet. 17 percent of the teachers do no use Internet at all in the class room.

When we look at educational tools and services that require broadband connection, we see the following image (Table 13.8). A lot of those 'real' broadband tools and services are not used at all. For example, videoconferencing is not used by 98 percent, complete productions, such as documentaries via Internet is not used by 88 percent, TV via Internet is not used by 86 percent, playing online games is not used by 81 percent, chat and MSN services are not used by 80 percent, online collaboration tools are not used by 79 percent and online audio material for music and language lessons is not used by 73 percent. A lot of those tools and services are familiar to the respondents though.

From the interviews we have come to learn that although many schools do not use video applications, the schools that do work with it are very enthusiastic. Applications in which video were used were especially popular with students as well as teachers. Video sessions with experts, such as a space travel expert or an America expert, or with students from other countries made a lot of impact on the students and teachers.

Table 13.8. Usage of broadband services and tools

	No, and not familiar	No, but familiar	Yes, daily	Yes, weekly	Yes, monthly	Yes, yearly
Playing online educational games	35%	46%	-	4%	5%	10%
Complete productions such as documentaries via Internet	34%	54%	-	-	-	12%
Short video clips/ animation via Internet (e.g., Functioning of a motor or the heart)	18%	39%	-	4%	15%	24%
Online photos/drawings/maps	4%	16%	10%	15%	19%	37%
Online audio material (music, languages, interviews)	11%	62%	3%	3%	7%	15%
Videoconferencing	22%	76%	-	1%	-	1%
Chat / MSN	-	80%	5%	5%	4%	5%
Online collaboration tools, such as electronic learning environments	41%	38%	8%	3%	5%	5%
TV via Internet	16%	70%	1%	-	5%	7%
Creating and provide educational broad-band content (e.g., student and teachers making educational videos)	28%	49%	1%	1%	4%	16%

In the future, the teachers want to use a variety of tools and services (Table 13.9). Teachers are a bit cautious and choosing mainly the services and tools that are already used. This may be caused by a lack of awareness. Short video clips and animation via Internet is most in demand (20 percent), followed by online photos/drawings/ maps (15 percent). But, more innovative things are (equally) wanted, such as online audio material (10 percent), complete productions such as online documentaries (10 percent), educational online games (10 percent), and online collaboration tools, such as electronic learning environments (10 percent).

Eighty-five percent of the teachers want to include and integrate Internet more and more in the classes. Apart from the very specific tools and services in the preceding passage, the teachers were asked what kinds of Internet applications they want to use in the future. The teachers want to use Internet for a variety of things (Table 13.4.): to look up information (30 percent), to show images and photos (28 percent), to hold examinations (20 percent), and to support interaction and discussion (19 percent).

Although teachers use the Internet in their classes and although they indicate that they want to use it more often in the future, there are some constraints, which prevent them from doing so (Table 13.10). The most important constraints are for using the Internet in class are lack of educational broadband content (or access to it) (26 percent), lack of necessary equipment (such as PCs) (22 percent), and lack of skills (19 percent).

Table 13.9. Which applications do teachers wish to use in the classes in the future?

Short video clips/ animation via Internet	20%
Online photos/drawings/maps	15%
Online audio material	10%
Complete productions such as documentaries	10%
Educational online games	10%
Online collaboration tools, such as electronic learning environments	10%
Creating and provide educational broadband content	9%
TV via Internet	7%
Videoconferencing	4%
Chat / MSN	3%
None	2%

Table 13.10. What are the impediments to use broadband in class (more answers allowed)

Lack of educational broadband content	26%
Lack of necessary equipment (such as PCs)	22%
Lack of skills	19%
The current infrastructure is sufficient, there is no need	14%
Impossible or hard to integrate in educational programme	13%
Unsuitable equipment (old PCs, etc.)	10%
There is no added value for the classes given	5%
No constraints	14%
Other	2%
Do not know	7%

Added Value

The added value of broadband for educational purposes in different terrains is acknowledged by the teachers, ICT coordinators, and school boards (Table 13.11). The most important added value of broadband in educational context is the possibility for internal educational applications or teaching material (53 percent). Also shared use and management of services and applications (28 percent) and external communication with students (25 percent) are mentioned as important benefits. External communication with parents, for example via Websites, is not as important as communication with the students, but is also mentioned in 16 percent of the cases. Support in processes such as administration is also an advantage of broadband (21 percent), but that does not automatically lead to cost reductions according to the respondents. Cost reductions are not seen as a real advantage of broadband in education (2 percent).

As can be seen from Table 13.12, teachers find broadband applications especially suitable for facilitating individual learning (23 percent), facilitating and stimulating location independent learning (20 percent), supporting the development of individual skills (16 percent), facilitating class activities (14 percent), and stimulating and facilitating collaboration between students (13 percent). Stimulating and facilitating interaction between teacher and student is a somewhat less important benefit of broadband.

Table 13.11 Added value of broadband in education? (max. 3 answers)

Internal educational applications (teaching material)	53%
Shared use and management of services and applications	28%
External communication with students (e.g. via email)	25%
Internal process support (e.g. administration)	21%
External communication with parents (e.g. via website)	16%
Cost reductions	2%
Other	2%
Do not know	8%

Table 13.12. Broadband applications are useful for ...(max. 3 answers allowed)

Facilitating individual learning	23%
Facilitating and stimulating location independent learning	20%
Supporting the development of individual skills	16%
Facilitating class activities	14%
Stimulating and facilitating collaboration between students	13%
Stimulating and facilitating interaction between teacher and student	8%
Do not know	4%
Other	2%

Furthermore, using multimedia educational content makes classes more interesting (24 percent), more versatile (24 percent), more fun (23 percent), and more effective (21 percent). According to the respondents it does not really make classes easier (4 percent) and more personal (3 percent) for the student.

Interview results show that the added value is also in students being able to learn in a more natural way in which the information had more impact and in which the students also remembered information for a longer period of time. Teachers even discover new capabilities of the students, such as taking on different roles in the project team, that otherwise would never have been detected.

Summary

The following tables summarize the most important added value of broadband for education and the thresholds for adoption in schools (Table 13.13) and in the class room (Table 13.14).

Discussion and Conclusion

In this section first, we reflect on the research, its results, limitations, implications, and we suggest further research. This research is one of the first to ask teachers,

*Table 13.13. Most important added values and thresholds for adopting broadband in the **school** (school boards and ICT managers)*

Thresholds	Willingness	No need	Current infrastructure is sufficient
	Ability	Financial Barrier	High subscription costs
Added value			Internal educational applications (better and variation in teaching material)
			Shared use and management of ICT services and applications
			External communication with students (e.g., via e-mail)

Table 13.14. Thresholds for adopting broadband in the classroom/classroom related activities (teachers)

Thresholds		Physical barrier	Lack of educational broadband content
	Ability	Technical barriers	Lack of necessary equipment (such as PCs)
		Cognitive barrier	Lack of skills
	Willingness	No need	The current infrastructure is sufficient, there is no need
Added value			Facilitating individual learning
			Facilitating and stimulating location independent learning
			Supporting the development of individual skills

school boards, and ICT-coordinators about their usage of and experiences with broadband. In this way, this research is innovative and also explorative. Due to the longitudinal design, the next measurements in the following years will provide us with more information on the further development of broadband in education. By repeated measurements, we will gain extra insights into the uses and impediments of broadband in education. With this information it becomes possible to not only construct a conceptual model that shows and predicts the adoption and usage of broadband Internet in education, but also of other new technologies in education.

The information provided by this research can help ISP and educational software suppliers determine what demands from the education area have not been met by the market. Also, the impediments that the respondents state can be addressed by various parties. But, since the majority of the respondents do not know what the impediments to switching to a new infrastructure are, could mean that the respondents are not involved in the decision making process or that they are not really aware of broadband and the possible impediments. This implies that more in-depth research should be done on this matter.

Because of the way this research is conducted, with an online questionnaire, it can be argued that especially respondents with a special interest for Internet have responded. In a future research a larger sample and a postal questionnaire or a combination of both (while minding the validity and reliability aspects of that) might lead to a more representative image. Also, a comparison between Dutch figures and other countries would be interesting. By doing so countries are able to get insights into the differences and later adopting countries can even learn from experiences in earlier adopting countries.

The Internet is frequently used in education at the moment, but the 'real' broadband applications are not used very much. Not even in schools that have fast connections such as most of the schools in this research. The good news is that teachers are interested in using applications in the future, such as: video, images, audio, games, and online collaboration tools. This positive attitude and intention to use is plausibly connected to the fact that the all those involved (teachers, ICT-coordinators, and school boards) are very aware of the added value of broadband in education. According to the teachers, ICT coordinators, and school boards broadband facilitates individual learning, location independent learning, and supports the development of individual skills. Also, they find it useful for internal educational applications (teaching material), shared use and management of services, and applications and external communication with students.

Some bottlenecks have to be solved though, before extensive use of broadband in the educational environment is a fact. Basically two conditions have to be met: (1) a school has to have access to broadband infrastructure; (2) once the infrastructure is there, it has to be used in the classroom.

The most important reason for not switching to a new broadband connection is that the current infrastructure is sufficient. This is only logical since teachers do not yet use tools and services that require high speed broadband. Of course, this is a matter of a causality dilemma: without infrastructure, there are no services and without services there is no need for infrastructure.

Another important impediment to switching to broadband is formed by the costs for subscription and for required devices and material. Here, the solution could be in combining forces and demand. Different initiatives in the Netherlands show that when schools combine demand for broadband roll out with local authorities and institutions such as libraries, the subscription costs decrease to levels that every school can afford (because infrastructure providers can adjust their offers to large scale demand).

The second condition for success of broadband in the educational environment is that the tools and services that the infrastructure offers, has to be used in the class-room. In the opinion of teachers there are some reasons why it is not used yet. First of all, there is a lack of educational broadband content. And here, again, we are confronted with the causality dilemma. In particular, (educational) publishers need to make the step towards digital supply and services, which implies adjustment of existing "print-oriented" business cases. Since the involved parties (schools) are aware of the added value and have intentions to use the 'real' broadband services and tools, it can be argued that it is very likely that there will be a mutual shaping. Once the infrastructure extends to more schools, new ideas for tools and services will rise, and the need for more bandwidth will be obvious for other schools. Both the development of broadband services and tools as well as the roll out should be pursued.

To further stimulate the use of broadband services and tools in the classroom, the lack of skills under teachers should be addressed. This lack of skills is now one of the most important constraints for teachers to use the services and tools. Also, there should be more awareness under school boards, because teachers may not always have a say in the decision making process.

In summary: teachers, ICT-coordinators, and school boards are interested in using broadband in their schools as they see the added value, but there seems to be an impasse. Schools can break out of the causality dilemma by giving an impulse to the market by combining forces and demand. Moreover, teachers need to be trained in using the new tools and services.

Acknowledgment

This research was conducted for Nederland BreedbandLand, SURFnet and Kennisnet. The original report is in Dutch (Dialogic, 2005b).

References

Auster, E., & Choo, C. W. (1993). Environmental scanning by CEOs in two canadian industries. *Journal of the American Society for Information Science*, *44*(4), 194-203.

Bouwman, H., Nouwens, J., Baaijens, J., & Slootman, A. (1996,in Dutch). *Een Schat aan Informatie Toegankelijkheid van Overheidsinformatie* (An Information Treasure: Accessibility of Government Information). Den Haag/Amsterdam: Rathenau Instituut/Otto Cramwinckel.

Broadband Stakeholder Group (2003). *Opportunities and barriers to the use of broadband in education. Report and strategic recommendations.* Retrieved January 6, 2006, from Broadband Stakeholder Group at http://www.broadbanduk.org

Daft, R.L., & Lengel, R.H. (1986). Organizational information requirements, media richness and structural design. *Management Science, 32*(5), 554-571.

Davis, F.D. (1989). Perceived usefulness, perceived ease of use, and user acceptance of information technology. *MIS Quarterly, 13*(3), 319-341.

Dialogic. (2005a, in Dutch). *Breedband en de Gebruiker 2004/2005.* Utrecht: Dialogic.

Dialogic. (2005b,in Dutch). *Breedbandmonitor Onderwijs.* Utrecht: Dialogic.

Duchastel, P.C., & Waller, R. (1979). Pictorial illustration in instructional texts. *Educational Technology, 19*(11), 20-25.

Goodyear, P., & Steeples, C. (1998). Creating shareable representations of practice. *Advance Learning Technology Journal, 6*(3), 16-23.

Fulk, J., Schmitz, J., & Steinfield, C. (1990). A social influence model of technology use. In J. Fulk & C. Steinfield (Eds.), *Organisations and communication technology* (pp. 117-140). Newbury Park: Sage.

ISP wijzer: www.ispwijzer.nl

ITS/IVA (2005, in Dutch). *ICT in cijfers, ICT-onderwijsmonitor studiejaar 2004-2005.* Nijmegen/Tilburg: ITS/IVA.

Massy, W. F., & Zemsky, R. (1995). *Using information technology to enhance academic productivity*. Washington, DC: Educom.

Mayer, R.E., & Gallini, J.K. (1990). When is an illustration worth ten thousand words? *Journal of Educational Psychology, 82*(6), 715-726

Niemi, J. A., & Gooler, D. D. (Eds.). (1987). *Technologies for learning outside the classroom. New directions for continuing education.* San Francisco: Jossey-Bass.

NOIE (National Office for the Information Economy). (2002). *Broadband in education: availability, initiatives and issues*, Canberra. Retrieved January 6, 2006, from www.dest.gov.au

Organization for Economic Cooperation and Development. The development of broadband access in OECD countries. October 29, 2005.

Pahad, A. (1998). Video films on acquired immune deficiency syndrome and its impact on cognitive and affective domains of college students. *Interaction, 17*(1), 64-71.

Underwood, J., Ault, A., Banyard, P., Durbin, C., Hayes, M., Selwood, I., et al. (n.d.). *Connecting with broadband: Becta sponsored pilot investigation of broadband technology impacts in schools: Literature review.* Retrieved January 24, 2006, from http://www.becta.org.uk

Rice, R.E. (1993). Media appropriateness: Using social presence theory to compare traditional and new organizational media. *Human Communication Research, 19*(4), 451-484.

Roden, S. (1991). Multimedia: The future of training. *Multimedia Solutions, 5*(1), 17-19.

Rogers, E.M. (1995). *Diffusion of innovations*. New York: Free Press

Shepard, R.N., & Cooper, L.A. (1982). *Mental images and their transformations.* Cambridge, MA: MIT Press/Bradford Books..

Short, J., Williams, E., & Christie, B. (1976). *The social psychology of telecommunications.* London: Wiley

Sitkin, S.B., Sutcliffe, K.M., & Barrios-Cholin, J.R. (n.d.). A dual capacity model of communication media choice in organizations. *Human Communication Research, 18*(4), 563-598.

Trevino, L. K., Daft, R. L., & Lengel, R. H. (1990). Understanding managers' media choices: A symbolic interactionist perspective. In J. Fulk & C. Steinfield (Eds.), *Organisations and communication technology* (pp. 71-94). Newbury Park, CA: Sage.

Van Dijk, J.A.G.M. (1996, in Dutch). *De netwerkmaatschappij: Sociale aspecten van nieuwe media*. Houten:Bohn Stafleu Van Loghum.

Van de Wijngaert, L. (2001). Internet in Context: Fysieke en affectieve toegang, geschiktheid; vraag, aanbod en context. In H. Bouwman (Ed.), *Communicatie in de Informatiesamenleving* (pp. 51-70). Utrecht, The Netherlands: Lemma.

Van de Wijngaert, L. (2004). Old and new media: A threshold model of technology use. In H. van Oostendorp, L. Breure, & A. Dillon (Eds.), *Creation, use, and deployment of digital information* (pp. 247-261). Mahwah, NJ: Lawrence Erlbaum Associates.

Van de Wijngaert, L., Vermaas, K., & Maltha, S. (2003). Broadband technology and services from a user perspective. In L. Haddon, E. Mante-Meijer, B. Sapio, K.H. Kommonen, L. Fortunati, & A. Kant (Eds.), *Proceedings of The Good The Bad and the Irrelevant* (pp. 179-185). Helsinki.

Vermaas, K. (2005,in Dutch). Een supersnelle internetverbinding. *Computers op School* (10/17), Groningen: ESS.

Division 3

Developing Country
Perspectives and Implications

Chapter 14

Factors Affecting Consumer Adoption of Broadband in Developing Countries

Abstract

This chapter empirically examines factors affecting the adoption of broadband in the developing countries of Bangladesh and the Kingdom of Saudi Arabia (KSA). In the case of Bangladesh, attitudinal, normative, and control factors—discussed in the UK case study in Division I of this book—were used and adapted in order to provide insights about broadband adopters and non-adopters within the developing nations. In order to examine the adoption of broadband in the KSA, a number of variables were employed, which also included some of the variables discussed in the UK case study in Division I. As the Internet was introduced comparatively late in Bangladesh (in 1996), in early 2004 the total penetration of Internet within the country was only 0.25 percent (Totel, 2004). It was suggested that the major obstacles associated with low Internet penetration were the low economic status and still-developing infrastructure within the country (Totel, 2004). A recent media report further emphasised that "Bangladesh is not anywhere on the global broadband map, but it is doing its best to get online. Local service provider, DNS SatComm has started deploying fixed wireless gear from Cambridge Broadband and will offer access to government offices, and other commercial entities" (Malik, 2005). It has also been suggested that Internet connection is slow and costly and not affordable by the general public (Hossain, 2004). Given the situations of Bangladesh in terms of demography, telecommunication infrastructure, and affordability of Internet by people, it was felt that understanding factors including cost of Internet access and subscription affecting consumer adoption might help to encourage further diffusion of high speed Internet. In the KSA, the Internet has taken some time to diffuse and is therefore seen as a relatively new technology. The KSA first started with dial up connections and then moved on to adopt broadband and satellite connections to provide better data communication services to its citizens. However, even with the availability of broadband technology, the rate of adoption is considered to be relatively poor in comparison to other developed countries such as the UK, as well as newly industrialised leading broadband users, such as South Korea (Oh et al., 2003). This poor connectivity is often claimed to be caused by website filtration in the region. Consequently, broadband adoption has been slower than expected in the region. Furthermore, a survey of existing literature on broadband adoption suggests that although both macro and micro level studies were conducted in order to understand the deployment of broadband in the developed world and leading countries such as South Korea, none of these studies focus upon developing countries, such as Bangladesh and the KSA. Although this could be attributed to the slow infrastructure development and low rate of adoption within the two countries, this has provided the motivation for undertaking exploratory research in order to develop an understanding of the perceptions of consumers regarding broadband adoption in these developing nations. Thus, this chapter aims to explore the reasons for the slow adoption of broadband in Bangladesh and the KSA by examining the individual level factors affecting broadband uptake in both cases. The research will thereby seek to adapt the individual level factors from the UK case study (Division I) and attempt to examine if and why the adapted factors affect consumers' attitudes towards the adoption of broadband in the countries. The chapter begins with a brief discussion of the theoretical basis and variables employed to examine broadband adoption. This is followed by a brief discussion of the utilised research methods. The findings are then presented and discussed. Finally, a conclusion to the chapter is provided.

Theoretical Basis

The following theoretical constructs were adapted from the UK case study in order to examine broadband adoption in Bangladesh. It was proposed that the behavioural intentions (BI) to adopt broadband in Bangladesh will be determined by the following three types of constructs: (1) attitudinal constructs (*relative advantage, utilitarian outcomes, and hedonic outcomes*), which represent the consumers' favourable or unfavourable evaluation of the behaviour in question (i.e., adoption of broadband) (Rogers, 1995; Venkatesh & Brown, 2001); (2) normative constructs (*primary and secondary influence*), which represent the perceived social pressure to perform the behaviour in question (i.e., adoption of broadband) (Venkatesh & Brown, 2001), and (3) control constructs (*skills and facilitating conditions resources*), which represent the perceived control over the personal or external factors that may facilitate or constrain the behavioural performance (Venkatesh & Brown, 2001). Findings from the UK case study suggested that the majority of the examined constructs significantly influenced the BI to adopt broadband in UK households. Therefore, similar constructs have also been adapted to examine broadband adoption in Bangladesh.

The KSA study considered that the following constructs may influence citizens' attitudes towards broadband adoption in the context of the KSA: usefulness, resources, skills, service quality, compatibility, relative advantage of technology, usage socio-cultural factors (for example language and regulation through filtration), and demographic variables such as age, gender, education, occupation and income, type of accommodation, and whether one works from home or not. The majority of the constructs that were included in the case of the KSA were adapted from the UK case study, however, some constructs such as type of accomodation were identified from other relevent literature.

Research Methodology

A self-administered questionnaire with multiple-type closed questions was considered to be the primary survey instrument for data collection in both the cases. Due to the uncertainty regarding personnel using the broadband facility, the snowball or chain sampling (Fridah, 2002) method was adopted when selecting the respondents. In order to identify the first few respondents with Internet connection, researchers approached friends and colleagues who possessed the broadband connections at home using e-mail in order to complete the questionnaire. The respondents were also requested to recommend friends and family who had Internet connections at home and who may wish to participate in the research.

In the case of Bangladesh, the final questionnaire consisted of a total of 12 close-ended, multiple, Likert scale type questions. The final questionnaire was administered to a total of 80 broadband users in Bangladesh via email during August and October, 2005. All the respondents who replied were located in Dhaka, Bangladesh. Of the 80 questionnaires administered, 70 respondents returned the completed questionnaire, thus yielding a response rate of 87.5 percent.

In the KSA study, the final questionnaire consisted of a total of 18 questions that included close-ended, multiple and Likert scale type questions. The questionnaire was administered to a total of 150 broadband users via email during July and August, 2005. The majority of respondents who replied were located in two of the big cities of the KSA which are Riyadh, the capital, and Jeddah. Of the 150 questionnaires administered, 138 respondents returned the completed questionnaire, thus yielding a response rate of 92.0 percent.

The initial stage of data analysis involved checking the responses, and providing a unique identification number to each response. Using the SPSS application, descriptive statistics (i.e., frequencies, percentage and tables) were generated and reliability tests and regression analysis were conducted to analyse and present the research data obtained from the questionnaire.

Findings: Broadband Adoption in Bangladesh

Respondents' Profiles

Of the 70 respondents, only 12.9 percent represented the adopters of broadband and the remaining 87.1 percent were the non-adopters. The non-adopters of broadband included respondents accessing the Internet using narrowband (dial-up) at home and those who did not have Internet access at all. Of the 87.1 percent non-adopters category, 70 percent possessed a narrowband connection and 24.3 percent stated that they did not have any means of Internet access at home.

Regression Analysis

Influence of Relative Advantage, Utilitarian Outcomes, and Hedonic Outcomes on Attitudes of Consumers

The regression analysis was performed with attitude as the dependent variable and relative advantage, utilitarian outcomes and hedonic outcomes as the predictor variables. A total of 70 cases were analysed. From the analysis, a significant model emerged (F (3, 70) = 24.567, $p < .001$). The adjusted R square was 0.506. From the three independent variables only one variable, relative advantage, ($\beta = .479$, $p < .001$) was found to be significant. Both utilitarian outcomes ($\beta = .155$, $p = .204$) and hedonic outcomes ($\beta = .195$, $p = .131$) as predictor variables were found to be not significant.

Influence of Attitude, Primary Influence, Secondary Influence, Facilitating Conditions Resources, and Skills on Behavioural Intentions

The regression analysis was performed with behavioural intentions as the dependent variable and attitude, primary influence, secondary influence, facilitating conditions resources and skills as the predictor variables. A total of 70 cases were analysed. From the analysis, a significant model emerged (F (5, 70) = 18.239, $p < .001$). The adjusted R square was 0.555. Four predictor variables included in the analysis were found to be significant. These included facilitating conditions resources (β = .244, p = .009), attitude (β = .518, $p < .001$), primary influence (β = .369, p = .001) and secondary influence (β = .326, p = .002). Only skill as a predictor variable was not found to be significant (β = .175, p = .100). The size of β suggests that the attitude construct has the largest impact in the explanation of variations of behavioural intentions. This is followed by the primary influence construct and then secondary influence. The facilitating condition resources construct from the control category contributed the fourth largest variance of behavioural intentions.

Influence of Behavioural Intentions and Control Factors on Adoption Behaviour

A logistic regression analysis was performed with broadband adoption as the dependent variable and behavioural intention and facilitating conditions resources as the predictor variables. The dependent variable, which measures the broadband adoption behaviour, was categorical in nature and represented by Yes and No. Yes is equal to 1 if the respondent possesses broadband and 0 if they do not have broadband. A total of 70 cases were analysed and the full model was considered to be significantly reliable (χ^2 (2, N = 70) = 13.082, p = .001). This model accounted for between 17.3 percent and 33.7 percent of the variance in broadband adoption, and overall, 91.3 percent of the predictions were accurate. Both the behavioural intentions and facilitating conditions resources reliably predicted broadband adoption. The values of the coefficients reveal that each unit increases in behavioural intentions and the facilitating conditions resources score is associated with an increase in the odds of broadband adoption by a factor of .192 and 2.998 respectively. This means that facilitating conditions resources has a larger part in explaining actual adoption than behavioural intentions.

Findings: Broadband Adoption in the KSA

Respondents' Profiles

Of the 138 completed questionnaires that were returned, two questionnaires were discarded because they were not completed in full. This meant that a final sample of 136 questionnaires was used for all subsequent analysis. Of the 136 respondents, 92 (68 percent) used broadband, while 32 percent did not use broadband, although all respondents had knowledge about broadband technology. This meant that approximately 32 percent of Internet users still used dial up as opposed to broadband. With regard to gender, respondents were fairly well represented from both male and female categories, with females accounting for 44 percent and males comprising 56 percent of the respondents. The majority of the respondents were aged between 18 years and 50 years. With regards to education, only 5 percent were educated to secondary school level, while 57 percent were graduates and 38 percent had postgraduate levels of education. The usage level (number of hours on the Internet per day) was normally distributed and the majority of respondents spent on average 3 hours per day on the Internet. Of these, 72 percent also worked from home, whilst the remaining 28 percent only worked away from home (i.e., in an office environment). About 23 percent of respondents used the Internet for personal use, whilst the remaining 77 percent used it for both business and personal use. Of the respondents, 67 percent lived in houses and 33 percent lived in flats. With regards to the connection type, 55 percent used broadband, 12 percent used satellite and 32 percent still used dial up. This means that 68 percent respondents were connected to the Internet via high-speed connections.

Regression Analysis

A regression analysis was conducted with attitude towards broadband adoption as the dependent variable and usefulness, skills, resources, compatibility, technology, service quality, social/cultural, and control factors as predictor variables. A total of 136 cases were analysed. From the analysis, a significant model emerged (F (16, 136) = 4.544, p < .001). The R^2 for this analysis was found to be 0.396. This means that the factors explained 40 percent of the changes in the attitude. The R^2 of 0.4 (40 percent) is considered as a good value for a cross-sectional data involving many predictor variables. The overall regression analysis results suggest that out of the variables that were identified in the broad literature as having an impact on attitude towards broadband adoption in the context of the KSA, the significant variables were: the perceived usefulness of broadband, the service quality, the age of the respondent, the type of connection, and the type of accommodation. The remaining variables were found to have insignificant or no impact on the dependent variable. The following factors have a significant impact on consumers' attitude towards adoption of broadband: usefulness (β= .205, p = .015), service quality (β = .20, p = .036), age (β = -.384, p < .001, usage (β = .213, p = .009), connection type (β = .220, p = .010), and type of house (β = -.204, p = .013). However, the socio-cultural factor was found to be insignificant. This indicates that while the inhabitants of the KSA feel that there is some level of control of broadband by the authorities (the state), they do not feel that this control affects their attitudes towards adopting broadband technology.

This research presented on broadband adoption in the Kingdom of Saudi Arabia' established that the key factors that affected citizens' attitude towards broadband included usefulness, service quality, age, usage, type of connection, and accommodation type. The following is an attempt to provide explanations for the key findings.

The usefulness construct of a technology is important in the decision process, and findings clearly indicate that the KSA consumers considered this factor important. It may be an indication that the service provider is not creating enough awareness of the products and services that broadband can offer. Also, the findings provide some evidence that the KSA broadband service provider is not satisfying customers with the service quality they provide. A possible explanation for this poor service is the lack of competition since the KSA has only a single broadband provider. Early studies carried out in the UK on the broadband market suggest that, possibly, the lack of competition caused by incumbent monopolist of the market was leading to poor quality service, and also that consumers were not getting value for money (Choudrie & Lee, 2004). On the other hand, studies carried out in South Korea where broadband penetration was most successful in the world (Oh et al., 2003), showed that the quality of service was excellent. This indicates that high quality of service has a highly influential impact on the mass adoption of broadband technology. Therefore, it can be argued that the situation in the KSA could be mirroring that in the UK in comparison to South Korea, where services were seen to be of high quality.

In South Korea, the high level of broadband penetration was partly attributed to the level of competition and also the general level of education that broadband users possessed. It was also found that the demographic variable 'type of accommodation' significantly influenced the attitude towards broadband adoption: those who live in flats were more likely to adopt broadband compared to those who live in houses. Although it is not entirely clear why this is the case, a possible explanation can be that often the young and more educated are likely to be city workers and therefore more likely to live in flats. Hence, they are more likely to adopt a new innovation than, for instance, citizens who live in houses in more rural areas. Another important demographic variable which has an impact on the attitude to adopt broadband is the usage factor. However, many of the demographic variables did not have an impact on broadband adoption attitudes.

With regard to the insignificant constructs such as skills, resources, compatibility, technology, and socio-culture, there could be some possible explanations for these. For skills, it may be possible that broadband is perceived as an extension of the Internet, and consumers are very familiar with the technology and are confident that they can learn this as they adopt the technology. This is an indication that broadband technology may not be as innovative as we initially perceived and projected, but just an extension of a technology that consumers are familiar with.

With regard to resources, the findings show that these have not adversely affected individuals' attitudes to adopt broadband. While price was identified as a potential deterrent, in the KSA this does not seem to be the case. This is not surprising given the comments by Crabtree (2003) that the price of broadband has fallen substantially since its introduction, and this should provide some stimulus for its uptake. Crabtree (2003) argues that there are many micro barriers preventing consumers from migrating to broadband. However, the assumption here is that there should be a natural shift from dial up to broadband and this assumption needs to be tested and research conducted in order to find out why dial up users may not migrate to broadband. Indeed, the KSA still has 32 percent of Internet users on dial up who have not migrated to broadband although they can, and resources are not the limiting factor in this case. This means that these consumers may need to be convinced of the benefits of broadband to encourage them to migrate to broadband.

The findings on socio-cultural factors were unanticipated, because the prediction was that these factors have a negative influence on broadband adoption. The previous literature also identified that the South Korean culture in general was a critical factor in the success of broadband adoption (Oh et al., 2003). Given the strong traditional and conservative nature of the KSA culture, and the information that was gathered during brief interviews with some of the survey respondents, there were strong and persuasive arguments to predict that these factors have a negative impact on broadband adoption attitude. A possible explanation for the lack of impact by socio-cultural factors could be that most KSA citizens may have in fact learnt to live with the problems imposed by their socio-cultural background. While they are aware of regulations, most KSA citizens are determined not to let this affect their attitudes towards the adoption of broadband—they will use the permissible aspects of a technology to their best ability. In this context, it is possible to suggest that most consumers are convinced that the benefits of broadband far outweigh the limitations caused by regulations, such as filtration of Internet content. Another explanation could be that some consumers see filtration as a necessity to protect their socio-cultural values and their children in particular from potentially harmful Internet content. However, the extent to which this is exercised raises important issues for the government in being more upfront regarding what exactly is filtered.

This, coupled with an effective publicity campaign by the government, will no doubt encourage further uptake of broadband in the KSA.

Conclusion

This chapter examined empirically the factors affecting the adoption of broadband Internet in the developing countries of Bangladesh and the KSA.

The following main conclusions are drawn from the case from Bangladesh. A total of five constructs (attitude, primary influence, secondary influence, skills, and facilitating conditions resources) were expected to be correlated to the behavioural intentions of consumers when adopting broadband in Bangladesh. Of these five constructs all, bar skills, significantly correlated to the BI of consumers. In terms of the size of the effect of the four constructs that contributed significantly to the behavioural intentions of consumers—attitude, primary influence, secondary influence and facilitating conditions resources—attitude exhibited the largest and facilitating conditions resources demonstrated the least variance to the behavioural intentions of consumers when adopting broadband in Bangladesh. Primary influence explained the second largest variance, which was followed by secondary influence. Both behavioural intentions and the facilitating conditions resources significantly correlated to the Internet adoption behaviour. In terms of the relative impact of the two aforementioned constructs that contributed significantly to the broadband adoption behaviour, facilitating conditions resources had much higher impacts than behavioural intentions.

The KSA study found that the factors that had a significant influence on the adoption of broadband were: usefulness, service quality, age of consumer, connection type, and accommodation type. The research established that the service quality provided by the monopolist provider in the KSA was having a negative impact on the adoption of broadband. The research also established that the socio-cultural factors, although being important for consumers, did not negatively affect the adoption of broadband in the KSA.

As broadband technologies enable a range of communication and Internet services, studying individuals from Bangladesh and the KSA provides a useful starting point for understanding the adoption of broadband in developing countries. This research presents one of the initial efforts towards understanding the adoption behaviour of Internet consumers from the perspective of developing nations. The findings are specifically useful for ISPs and policy makers of Bangladesh and the KSA. Factors that are reported significant are of utmost importance and require attention in order to encourage further adoption and usage of Internet in both countries. Although the infrastructure problem is the most predominant in developing countries when it comes to broadband adoption, the usage patterns are also important as the technologies are gradually put in place. Additionally, the cost of using the traditional telephone network is very high so broadband Internet can be used as a replacement for offering communication services such as instant messaging or IP telephony. The important implications from both the studies are discussed in the next chapter (Chapter 15).

The first limitation of both the cases (Bangladesh and the KSA) is the generalisation of the findings, which is highlighted below. The generalisation of this study required collecting the random data from across Bangladesh and the KSA. Furthermore, this research intended to supplement the questionnaire data with interviews, which was not possible due to the shortage of time and resources: the data for this research has subsequently been collected within a short period of time and provides a snapshot of respondents' behaviours at one point in time. This can, however, be expanded over a longer period of time in order to provide longitudinal data. This will then eliminate any variables that may have produced anomalies in the subsequent results.

Acknowledgment

The author wishes to thank and acknowledge the overall contribution made to this research by Dr. Vishanth Weerakkody and Mrs. Shatha Makki for administering the survey questionnaire in the Kingdom of Saudi Arabia and to Naureen Khan for administering the survey questionnaire in Bangladesh.

References

Choudrie, J., & Lee, H. (2004). Broadband development in South Korea: Institutional and cultural factor. *European Journal of Information Systems, 13*(2), 103-114.

Crabtree, J. (2003). *Fat pipes, connected people-rethinking broadband.* London. Retrieved March 30, 2004, from the *iSOCIETY Report* at http://www.theworkfoundation.com/pdf/1843730146.pdf

Fridah M. W. (n.d.). *Sampling in research.* Retrieved April 29, 2004, from http://trochim.human.cornell.edu/tutorial/mugo/tutorial.htm

Hossain, A. (2004). *Access to Internet Bangladesh perspective.* Retrieved November 15, 2005, from http://www.apdip.net/documents/evaluation/indicators/itu-bd16112004.ppt

Malik, O. (2005). *Bangladesh goes fixed wireless for broadband.* Retrieved October 10, 2005, from http://www.gigaom.com/2005/04/17/bangladesh-goes-fixed-wireless-for-broadband/.

Oh, S., Ahn, J., & Kim, B. (2003). Adoption of broadband internet in Korea: The role of experience in building attitude. *Journal of Information Technology, 18*(4), 267-280.

Rogers, E. M. (1995). *Diffusion of innovations.* New York: Free Press.

Totel. (2004). *Bangladesh-data, internet and broadband.* Retrieved November 15, 2005 at http://www.totel.com.au/asian-telecommunications-research.asp?toc=2289,

Venkatesh, V., & Brown, S. (2001). A longitudinal investigation of personal computers in homes: Adoption determinants and emerging challenges. *MIS Quarterly, 25*(1), 71-102.

Chapter 15

Implications and Future Trends

Abstract

This chapter presents the implications of the research discussed in this book and outlines future research trends in the area of consumer adoption and usage of broadband. The findings of the studies detailed in this book generate a number of implications that may be relevant to policy makers, Internet service providers (ISPs), and other relevant stakeholders for increasing consumers' adoption of broadband. The chapter begins by a discussion concerning the implications of this research for the government, followed by the implications for the Internet/broadband service providers and, ultimately, the implications for content providers and emerging electronic services. Finally, a discussion on the future trends in the area of broadband adoption and diffusion is provided.

Implications for the Government

Within the UK case study, self-efficacy was found to be a significant factor influencing individuals' behavioural intention to adopt broadband, which brings to the forefront policy-related issues. This suggests that there is a need to equip citizens with the skills to use computers and the Internet. When it comes to the government's role in equipping citizens, it is important to take a segmental approach for identifying and providing relevant skill-oriented courses to those citizens who do not have normal opportunities to learn and use the computer, Internet and other related emerging technologies and applications such as e-government and e-commerce. The findings related with demographic variables, such as age, were found to be negatively correlated in all case studies of this book including the UK one. This should be considered for identifying segments for attention in terms of skills and resources. Policy makers should ensure that older segments of society should be equipped with essential ICT skills. Ignoring such an issue will increase the digital divide between older and younger population of the digital society.

Governance of many countries including the UK are becoming more and more technology dependent as all local and central government services to citizens are transformed into electronic delivery medium. However, both government and citizens cannot realise the benefits of such a transformation until all citizens are equipped with the skill to access government services via an electronic medium. Citizens' ICT skills are of utmost important for survival and progress of the digital society. In the UK, the government provides free Internet access in libraries, however, not all people—particularly those in the older age categories—regularly visit the library so it may not prove very useful for motivating and encouraging such a segment of the population to learn and use the computer and Internet. An alternative strategy that may prove useful is providing Internet access in public places such as pubs and in old people's homes. It also require choosing and equipping one or two people with essential ICT skills who regularly visit pubs as a change agent who may then easily motivate others to learn and use such technologies.

It is clear from the case studies that the government has an important role to help both the broadband provider and the consumers. The implication for the government in terms of the broadband provider is to ensure that the company provides a better service quality. In a monopoly situation like this, it is often very tempting for the monopoly supplier to be inefficient. The government should therefore consider increasing the number of service providers and help create a more competitive market for broadband provision. The UK has done this in two ways in efforts to break down BT's monopoly of broadband provision. In the UK, the first action has been to allow cable providers, such as NTL and TeleWest, to develop the infrastructure and provide broadband services. This has worked well as quite a substantial proportion of consumers including public services such as schools and hospitals use

NTL/TeleWest broadband. Although BT still dominates the market, there has been a significant migration to NTL by both businesses and domestic consumers.

However, in many developing countries, including Bangladesh and the KSA, broadband is provided by only one supplier due to government regulation that appears not to encourage competition in this sector. The second action in the UK has been to 'force' BT to sell broadband at wholesale price to many local providers who use the BT system to run their own broadband. This has proved very popular with the emergence of companies, such as Wanadoo and VirginNet, supplying broadband at cheaper prices and tailoring packages to individuals depending on their need levels. For example, there are cheaper packages for light users (with a limit on download megabytes per month) to expensive high-speed services for those who download large volumes of content such as music and multimedia content. These tailored packages are proving to be very popular with consumers. This action has been encouraging as these companies are also offering other telephony services; this industry is growing rapidly resulting in a higher uptake rate of broadband. In contrast, in the KSA and many other developing countries, still there is no wholesale pricing of broadband by their sole supplier.

At the individual level, the government has to conduct a publicity campaign to make consumers more aware of the benefits of broadband. Of particular importance in the case of the KSA is the need for the government to address the issues relating to filtration. In the KSA, a significant proportion of the population is not happy with the regulation through filtration. The government has to therefore conduct a strong publicity campaign that will explain and educate consumers as to why filtration is necessary and highlight the long-term benefits of this action to the KSA society. There is evidence in literature (Lee et al., 2003; Udo, 2001) that parents fear for Internet security, especially when children can access and download adult material with less effort using broadband. The government should be more forthright when handling what exactly is filtered so that they can gain public confidence, which can ultimately lead to the increased adoption of broadband. Finally, the government's responsibility to increase awareness of the usefulness of broadband among citizens is also important particularly in relation to addressing the critical issue of the digital divide.

Implications for ISPs/Broadband Providers

As discussed before in the UK case study, ISPs or broadband service providers need to change their strategies continuously as customers' needs and demands are likely to evolve. Broadband price plans in the UK are still perceived as being expensive by consumers. The number of steps that have to be completed which includes signing the contract and installation fees—may also be acting as a barrier to subscribing to broadband. This suggests that ISPs should develop strategies that may eliminate signing the 12/18 months contract and provide free installation service for first time adopters of the technology. This can persuade consumers that they are not going to lose out and there is no harm in trying and, if they not satisfied with the service, then they are free to try the services of another provider. Also, there are consumers who cannot afford the current price plan. As a result, the ISPs may consider providing alternative price plans in order to create mass-market demand, which is an issue currently being emphasised. The ISPs may overcome the challenges described by offering differential price plans and segmenting specific broadband subscription packages. For example, the ISPs can differentiate within the offered subscription price ranges depending on factors, such as the income levels and needs of users. Also, with an increasing demand in the lower income segments and those with fewer needs of broadband, ISPs may offer price plans that can compete with the current price plan of un-metered narrowband. Currently, there is a price gap between the two packages, therefore a low price plan of un-metered narrowband is an inhibiting factor for broadband adoption in segments with lower incomes and fewer needs.

Since cost is not a factor of consideration when segmenting between the higher income and occupation levels, it should be offered in broadband packages with even faster speeds and appealing content. Such packages may assist in illustrating the clear benefits of broadband over narrowband to consumers of higher income and occupation levels, and provide them with added reasons for subscribing to broadband. For example, since relative advantage is found to be the strongest construct, it indicates that ISPs have to provide broadband services to consumers in such a package that would illustrate a clear advantage over narrowband consumers. Similarly, facilitating conditions resources is the third most important factor in terms of influencing BI to adopt broadband. This has implications for both ISPs and policy makers. For instance, ISPs have to think about more consumer-centric services and alternative price plans so that all consumers who want to subscribe to broadband are able to do so. Policy makers have to provide alternative places for broadband access where lower income groups or those who cannot afford it can access and use high speed Internet. This may help to increase behavioural intention to adopt broadband and therefore encourage overall adoption and diffusion of broadband within UK households.

Similar implications also emerged from the study of developing countries such as the KSA. The usefulness of broadband was found to have an important influence in the KSA. As such, the broadband provider needs to convince potential users of the benefits that they can gain from using broadband. To do this, the company has to continually produce and disseminate information about the benefits of broadband and the products and services that can be accessed using broadband. It is important for the company to realise that content cannot be separated from infrastructure matters, and therefore both need to continually develop. Also, the speed of broadband connections need to continually evolve like in the case of South Korea where broadband speeds are up to 10 times faster than those in the UK (Oh et al., 2003). Regrettably, the broadband speed is quite limited in the KSA and the KSA provider therefore has much to learn from the South Korean experience.

The service quality was also identified as being a significant factor in influencing the broadband adoption decision and continued adoption behaviour in both the UK and the KSA case studies. In the KSA study, the service quality was found significant, indicating that the provider has to improve service quality significantly. Although a monopoly can lead to inefficiency, an advantage of a monopoly is that the single provider can divert resources aimed at fighting off competition to improving the service quality. In this context, it may be necessary for the provider to seek the services of or partner with external organisations to provide high quality, value added services to consumers like in the case of BT in the UK. Although the UK broadband market now has a very high number of ISPs competing with each other, the UK case study suggests that some consumers are still unhappy with the services they are receiving from their providers. The study highlights that the quality of service received has a negative relationship with individuals' desire to continue subscribing to broadband with the same providers. This is an important consideration for ISPs to think and act upon: it is more difficult and expensive to attract and create a consumer base than retaining an existing consumer base. Thus, ISPs need to carefully listen to consumer complaints and try to solve their problems as soon as possible with minimum cost to these consumers. This is something currently lacking in the UK broadband market.

Furthermore, the KSA broadband provider also needs to target specific groups of people. First, there is evidence that younger people are more likely to adopt broadband compared to older people. Therefore, the provider needs to devise marketing strategies to target older people in order to increase migration to broadband. Secondly, the provider needs to target people who live in houses compared to those who live in flats, since this research indicates that those who live in flats are more likely to adopt broadband compared to those who live in houses. However, the provider may first need to conduct surveys to determine why this is the case. Finally, the provider needs to target low users (less than 2 hours on the Internet) as high users are potentially more likely to adopt broadband anyway.

Implications for the Content Providers

Since both utilitarian outcomes and hedonic outcomes were considered important factors for explaining behavioural intentions in the case studies discussed within this book, it is important to integrate more content and applications for the purpose of household and entertainment utility. South Korean broadband and content providers successfully utilised online games as a killer application/content for promoting broadband adoption. Similar strategies were, unsuccessfully, attempted to be applied in the UK. Cultural differences and dissimilarities in lifestyle between different countries require different strategies and content to promote broadband adoption. Findings from the UK case study suggest that online games were the least popular online activities undertaken by UK consumers at the time the survey was conducted. This clearly suggests that online games and gambling will not work as a killer application in the UK market. This necessitates adopting different strategies than those employed in South Korea. In the UK, it might be helpful if broadband providers and content providers develop a system for delivering real time sport events (i.e., soccer/football) and on-demand video/music to consumers with broadband connection.

The findings also have important implications for the electronic mass media and telecommunication industries. The mass-media industry is likely to benefit from the diffusion of broadband, as more respondents use online material rather than utilising traditional reading resources. This may encourage the online media industry to attract revenues from advertising and subscription fees. According to the findings, receiving, and making phone calls have decreased for both narrowband and broadband consumers. Therefore, the telecommunication industry may have to transform in terms of its business model. For example, the new business model may find a means of pricing consumers at instances of phone calls, which makes using VoIP or other emerging applications a consequence of a broadband environment.

Future Trends

This book presented the issue of broadband adoption from the consumer perspective. However, the problem of the slow rate of broadband adoption and usage in many countries includes factors that may not be covered only by examining consumer perspectives. Similarly, the impact of the adoption and use of broadband is briefly touched within this text and the following discussion sets a trend for future research.

In terms of a developing country perspective, both the KSA and Bangladesh studies suggest that there is a need to look regulation and infrastructure development issues at the national level. However, in terms of developed countries where progress has been made towards infrastructure development in urban areas, there is a need to determine the factors affecting broadband adoption and use in the rural and farm economic sectors and the development of broadband infrastructure in rural communities. Researchers within the developed countries context should systematically examine the broadband issues at various levels, for example, heterogeneous coverage, adoption, and usage of broadband. This is also linked with the issue of the digital divide, so future research efforts should be focused upon examining the varying levels of adoption in different sections of society and formulating strategies that policy makers may apply to create more homogeneity in the digital society.

Broadband is also considered a critical infrastructure and a measure of international competitiveness. Sometimes it also termed as the life blood for modern economy as it likely to affect many businesses, particularly small and medium enterprises (SMEs). However, like individual consumers, SMEs are also slow in adopting broadband. Factors that may affect individual consumers are likely to differ with those affecting the adoption of broadband by SMEs. This book provides an in-depth understanding of the factors affecting broadband adoption by consumers, but work towards understanding the factors affecting broadband adoption by SMEs remains largely untouched and therefore requires immediate attention by researchers.

New electronic services such as e-government services are currently being implemented in many countries. The diffusion and adoption of high speed Internet is a pre-requisite for the successful adoption of such emerging electronic services by citizens. This suggests that studying the impact of broadband on consumers, particularly in areas such as consumer adoption of new communication methods, music and software downloads, entertainment, retail, travel, and tourism on an individual basis can be beneficial in determining the real impact of broadband. Furthermore, there is a need to explore other associated issues such as the positive and negative impacts of these changes on the growth and development of the Internet, the diffusion and sustainability of broadband technology, family life and work, social interaction and development, and growth of the business-to-consumer, consumer-to-consumer electronic commerce, and e-government services areas. Widespread adoption of broadband is also likely to change the way many businesses undertake their business processes by affecting the value chain. Therefore, it is important to examine the business model of many sectors, particularly television, telecommunications, publishing and the picture/photo industry. For example, future research needs to examine the way the adoption of broadband will influence the economics of the motion picture industry. Also, with broadband facilitating the implementation of IPTV, the issue arises as to how IPTV will affect the current business model of the media and broadcasting/television industry. Similarly, broadband is an enabling technology for PC to PC communication such as VoIP, so a further issue which requires attention is how broadband VoIP will influence the telecommunications industry.

References

Lee, H., O'Keefe, B., & Yun, K. (2003). The growth of broadband and electronic commerce in South Korea: Contributing factors. *The Information Society*, *19*, 81-93.

Oh, S., Ahn, J., & Kim, B. (2003). Adoption of broadband Internet in Korea: The role of experience in building attitude. *Journal of Information Technology*, *18*(4), 267-280.

Udo, G. J. (2001). Privacy and security concerns as a major barrier of e-commerce: A survey study. *Information Management and Computer Security*, *9*(4), 165-174.

About the Contributors

Yogesh Kumar Dwivedi is a lecturer of information systems at the School of Business and Economics at the University of Wales Swansea, UK. He obtained his PhD, entitled 'Investigating consumer adoption, usage and impact of broadband: UK households,' and MSc in information systems from the School of Information Systems, Computing, and Mathematics at Brunel University, UK. He also holds a BSc (biology) from the University of Allahabad, India, and an MSc (plant genetic resources) from the Indian Agricultural Research Institute, Pusa Campus in New Delhi, India. His primary research interests focus upon the adoption and diffusion of information and communication technologies (ICTs) in organisations and society. In general, he is interested in investigating how consumers, organisations and society deploy and use ICTs for various purposes such as delivering education, e-commerce, e-business, and e-government services. He has co-authored more than 40 papers in academic journals and international conferences. He is a member of the editorial board/review board of the following journals: *Transforming Government: People, Process and Policy; Journal of Enterprise Information Management; International Journal of Electronic Finance; Journal of Computer Information Systems,* as well as being a guest/issue editor of the *Journal of Electronic Commerce Research.* He is a co-chair of three mini-tracks in the forthcoming American Conference on Information Systems, 2007. He is a member of the Association of Information Systems (AIS) and Life Member of the Global Institute of Flexible Systems Management, New Delhi.

* * *

Elizabeth Enabulele received a BSc (Hons) in mathematics (1990) from Obafemi Awolowo University, Ile-Ife in Nigeria and an MSc in distributed information systems (2003), from Brunel University, United Kingdom. She is currently a PhD researcher at School of Information Systems, Computing and Mathematics at Brunel University. Her research interests includes quality aspect of broadband connection, stakeholders and broadband, computer networking, quality of service.

Gheorghita Ghinea received a BSc and BSc (Hons) in computer science and mathematics (1993 and 1994, respectively), and an MSc in computer science (1996) from the University of the Witwatersrand, Johannesburg, South Africa. He then received a PhD in computer science from the University of Reading, United Kingdom (2000). He is a senior lecturer in the Department of Information Systems and Computing, Brunel University. His research interests span perpetual aspects of multimedia, quality of service and multimedia resource allocation, as well as computer networking and security issues.

Sven Maltha is an industrial economist. Since 1998 he has been partner/senior advisor and co-founder of Dialogic Innovation & Interaction, an independent Dutch research consultancy mainly active in the ICT and innovation domain (policy research and advice). His current research and advisory work are mostly at the crossroads of telecom, media, innovation, and competition policy. Most recent work focuses on broadband and covers also the convergence issue, competition between infrastructures, frequency policies, innovation in new media and user involvement in technology and service development. He worked for various Dutch ministries, provinces and Dutch municipalities as well as Dutch telcos and cable companies and the European Commission. In 2004, he was secretary of the Impulse Commission Broadband and advised the Minister of Economic Affairs (*Towards a National Strategy for Broadband*). In 2002, he was secretary of the National Broadband Expert Group, which advised the Cabinet on broadband policy.

John de Ridder is a former Telstra chief economist with the last 10 years at Telstra in retail and wholesale pricing roles. Over the last four years he has worked as a consultant to governments, international agencies, and private operators. Details of his activities can be found at www.deridder.com.au.

Karianne Vermaas is a researcher at the research-based consultancy firm Dialogic Innovation & Interaction in Utrecht, The Netherlands. At this moment she is conducting her PhD research at Utrecht University in the Department of Information and Computing Sciences. Her research focuses on the adoption and use of new technologies from a user perspective. This includes the use of (broadband) internet in the residential and educational context, for government information and services and for health information and services. She is for example involved in a longitudinal, multi-client research project on 'Broadband and the User.' Her research is commissioned by several ministries and (large) organisations, such as cable and telecom companies, the Dutch police and educational organisations.

Lidwien Van de Wijngaert is assistant professor at Utrecht University in the Department of Information and Computing Sciences. In a broad sense the research of Dr. van de Wijngaert addresses the adoption, implementation, uses, and effects of information technology both within the organisational as well as the domestic context. The goal of her research is to obtain insight into how information technology can effective and efficiently be used. The starting point is that this insight can best be obtained by maintaining a user perspective and matching user needs with technological capabilities. The research has a strong empirical basis and seeks collaboration with academia as well as industry and consumer organisations.

Index

A

asymmetric digital subscriber line (ADSL)
  2, 243, 263

B

Bartlett's test of sphericity 120
British Telecommunication (BT) 5
broadband 2–15
  adoption in Bangladesh and the Kingdom
    of Saudi Arabia (KSA) 285–295
  findings 288–292
  theoretical basis 286
  adoption studies 6, 138, 173
    adopters 139
    non-adopters 139
  attitudinal constructs 27, 173
    facilitating conditions resources (FCR)
      99
    hedonic outcomes (HO) 30, 99, 176
    relative advantage (RA) 29, 99, 175
    service quality (SQ) 31, 99, 177
    utilitarian outcomes (UO) 29, 99, 175
  conceptual model 22, 77

  description 25
  foundation 22
  control constructs 33, 179
    facilitating conditions resources
      34, 180
    knowledge (K) 35, 99, 181
    self-efficacy (SE) 35, 99, 180
  demographic variables 36, 184
    age 37, 141
    education 38, 143
    gender 37, 143
    occupation/income 39, 145
  dependent variables
    behavioural intentions (BI) 40
    behavioural intention to change the
      service provider (BISP) 99
    broadband adoption behaviour (BAB)
      40
  factor analysis 120
  in Dutch education 261–284
    accessibility 269
    classroom usage 273–275
    current subscriptions 263
    results 272

thresholds 268
value for education 267
instrument development process 78, 117–134
  confirmatory survey 117–134
  content validation (stage 2) 84
    findings 87
  exploratory survey (stage 1) 79
    findings 80
  findings 135–163
    respondents' profile 136
  instrument testing (stage 3) 90
  in The Netherlands (broadband vs. narrowband study) 241–260
    results 247–253
normative constructs 31, 178
  influences 32
    primary 178
    secondary 179
quality regulation 209–221
  framework 211
  research findings 215
    government 219
    price 219
    services 217
    usability 218
regression
  analysis I 147
  analysis II 147
  analysis III 150
  analysis IV 153
  logistic 151
research
  aims and objectives 8
  approach 10, 54
    non-response 64, 118
    response rates 66, 118
    survey 58–63
  contributions 11
  data analysis 69
  epistemology 52
  future trends 237, 301
  limitations 235
  main conclusions 227
  model of broadband adoption (MBA) 182

offerings and implications 229, 296–303
  for content providers 300
  for ISPs/broadband providers 298
  for the government 297
  to industry and policy 232
  to theory 230
  overview 223–227
  problem 4
  reliability test 119
Stakeholder Group (BSG) 2
usage and impact studies 6, 40–42, 153–162, 186–191

C

content validity ratio (CVR) 86, 130

D

decomposed theory of planned behaviour (DTPB) 20
Department for Work and Pensions (DWP) 201
digital literacy 76, 135, 222, 261

E

e-government services 197–208
  citizen adoption 200
  definition and benefits 198
  development in the UK 201
  government gateway 197–208
    citizens awareness and adoption 202
    demographic variables 203–206
electronic services 164–171
  current and future use of 164–171
    communications 165
    downloading 167
    e-commerce 169
    information producing 167
    information seeking 166
    media streaming 169
European Information Society (EIS) 199

I

integrated services digital network (ISDN) 243, 263

International Telecommunication Union
(ITU) 2

K

Kaiser-Meyer-Olkin (KMO) 120

M

model of adoption of technology in households (MATH) 20

N

non-quantitative positivist research (non-
QPR) 54

O

Organisation for Economic Co-operation
and Development (OECD) 2

P

perceived behavioural control (PBC) 33
principal component analysis (PCA) 120
public switched telephone network (PSTN)
263

Q

quantitative positivist research (QPR) 54

S

service level agreement (SLA) 210
sheep annual premium scheme application
(SAPS) 201

T

technology acceptance model (TAM)
17, 20
theory
diffusion of innovations (DI) 17, 18
of planned behaviour (TPB) 17, 19, 77
of reasoned action (TRA) 17

U

unified theory of acceptance and use of
technology (UTAUT) 18

use diffusion model (UD) 21

V

vocational education and training and adult
education (VET) 271